LONDON MATHEMATICAL SOCIETY LECTURE NOTE SERIES

Managing Editor: Professor I.M. James
Mathematical Institute, 24-29 St Gil

1. General cohomology theory and K-t
4. Algebraic topology, J.F.ADAMS
5. Commutative algebra, J.T.KNIGHT
8. Integration and harmonic analysis ... EDWARDS
9. Elliptic functions and elliptic cu... VAL
10. Numerical ranges II, F.F.BONSALL & J.DUNCAN
11. New developments in topology, G.SEGAL (ed.)
12. Symposium on complex analysis, Canterbury, 1973, J.CLUNIE & W.K.HAYMAN (eds.)
13. Combinatorics: Proceedings of the British Combinatorial Conference 1973, T.P.McDONOUGH & V.C.MAVRON (eds.)
15. An introduction to topological groups, P.J.HIGGINS
16. Topics in finite groups, T.M.GAGEN
17. Differential germs and catastrophes, Th.BROCKER & L.LANDER
18. A geometric approach to homology theory, S.BUONCRISTIANO, C.P. BOURKE & B.J.SANDERSON
20. Sheaf theory, B.R.TENNISON
21. Automatic continuity of linear operators, A.M.SINCLAIR
23. Parallelisms of complete designs, P.J.CAMERON
24. The topology of Stiefel manifolds, I.M.JAMES
25. Lie groups and compact groups, J.F.PRICE
26. Transformation groups: Proceedings of the conference in the University of Newcastle-upon-Tyne, August 1976, C.KOSNIOWSKI
27. Skew field constructions, P.M.COHN
28. Brownian motion, Hardy spaces and bounded mean oscillations, K.E.PETERSEN
29. Pontryagin duality and the structure of locally compact Abelian groups, S.A.MORRIS
30. Interaction models, N.L.BIGGS
31. Continuous crossed products and type III von Neumann algebras, A.VAN DAELE
32. Uniform algebras and Jensen measures, T.W.GAMELIN
33. Permutation groups and combinatorial structures, N.L.BIGGS & A.T.WHITE
34. Representation theory of Lie groups, M.F. ATIYAH et al.
35. Trace ideals and their applications, B.SIMON
36. Homological group theory, C.T.C.WALL (ed.)
37. Partially ordered rings and semi-algebraic geometry, G.W.BRUMFIEL
38. Surveys in combinatorics, B.BOLLOBAS (ed.)
39. Affine sets and affine groups, D.G.NORTHCOTT
40. Introduction to Hp spaces, P.J.KOOSIS
41. Theory and applications of Hopf bifurcation, B.D.HASSARD, N.D.KAZARINOFF & Y-H.WAN
42. Topics in the theory of group presentations, D.L.JOHNSON
43. Graphs, codes and designs, P.J.CAMERON & J.H.VAN LINT
44. Z/2-homotopy theory, M.C.CRABB
45. Recursion theory: its generalisations and applications, F.R.DRAKE & S.S.WAINER (eds.)
46. p-adic analysis: a short course on recent work, N.KOBLITZ
47. Coding the Universe, A.BELLER, R.JENSEN & P.WELCH
48. Low-dimensional topology, R.BROWN & T.L.THICKSTUN (eds.)

49. Finite geometries and designs, P.CAMERON, J.W.P.HIRSCHFELD & D.R.HUGHES (eds.)
50. Commutator calculus and groups of homotopy classes, H.J.BAUES
51. Synthetic differential geometry, A.KOCK
52. Combinatorics, H.N.V.TEMPERLEY (ed.)
53. Singularity theory, V.I.ARNOLD
54. Markov processes and related problems of analysis, E.B.DYNKIN
55. Ordered permutation groups, A.M.W.GLASS
56. Journées arithmétiques 1980, J.V.ARMITAGE (ed.)
57. Techniques of geometric topology, R.A.FENN
58. Singularities of smooth functions and maps, J.MARTINET
59. Applicable differential geometry, M.CRAMPIN & F.A.E.PIRANI
60. Integrable systems, S.P.NOVIKOV et al.
61. The core model, A.DODD
62. Economics for mathematicians, J.W.S.CASSELS
63. Continuous semigroups in Banach algebras, A.M.SINCLAIR
64. Basic concepts of enriched category theory, G.M.KELLY
65. Several complex variables and complex manifolds I, M.J.FIELD
66. Several complex variables and complex manifolds II, M.J.FIELD
67. Classification problems in ergodic theory, W.PARRY & S.TUNCEL
68. Complex algebraic surfaces, A.BEAUVILLE
69. Representation theory, I.M.GELFAND et. al.
70. Stochastic differential equations on manifolds, K.D.ELWORTHY
71. Groups - St Andrews 1981, C.M.CAMPBELL & E.F.ROBERTSON (eds.)
72. Commutative algebra: Durham 1981, R.Y.SHARP (ed.)
73. Riemann surfaces: a view toward several complex variables, A.T.HUCKLEBERRY
74. Symmetric designs: an algebraic approach, E.S.LANDER
75. New geometric splittings of classical knots (algebraic knots), L.SIEBENMANN & F.BONAHON
76. Linear differential operators, H.O.CORDES
77. Isolated singular points on complete intersections, E.J.N.LOOIJENGA
78. A primer on Riemann surfaces, A.F.BEARDON
79. Probability, statistics and analysis, J.F.C.KINGMAN & G.E.H.REUTER (eds.)
80. Introduction to the representation theory of compact and locally compact groups, A.ROBERT
81. Skew fields, P.K.DRAXL
82. Surveys in combinatorics: Invited papers for the 9th British Combinatorial Conference 1983, E.K.LLOYD (ed.)
83. Homogeneous structures on Riemannian manifolds, F.TRICERRI & L.VANHECKE
84. Finite group algebras and their modules, P.LANDROCK
85. Solitons, P.G.DRAZIN

London Mathematical Society Lecture Note Series. 85

Solitons

P.G. DRAZIN

Professor of Applied Mathematics, University of Bristol

CAMBRIDGE UNIVERSITY PRESS
Cambridge
London New York New Rochelle
Melbourne Sydney

CAMBRIDGE UNIVERSITY PRESS
Cambridge, New York, Melbourne, Madrid, Cape Town, Singapore, São Paulo

Cambridge University Press
The Edinburgh Building, Cambridge CB2 8RU, UK

Published in the United States of America by Cambridge University Press, New York

www.cambridge.org
Information on this title: www.cambridge.org/9780521274227

First published 1983
Re-issued in this digitally printed version 2008

A catalogue record for this publication is available from the British Library

Library of Congress Catalogue Card Number: 83–7170

ISBN 978-0-521-27422-7 paperback

To JUDITH

CONTENTS

PREFACE

The theory of solitons is attractive; it is wide and deep, and it is intrinsically beautiful. It is related to even more areas of mathematics and has even more applications to the physical sciences than the many which are indicated in this book. It has an interesting history and a promising future. Indeed, the work of Kruskal and his associates which gave us the inverse scattering transform is a major achievement of twentieth-century mathematics. Their work was stimulated by a physical problem and is also a classic example of how computational results may lead to the development of new mathematics, just as observational and experimental results have done since the time of Archimedes.

This book originated from lectures given to classes of mathematics honours students at the University of Bristol in their final year. The aim was to make the essence of the method of inverse scattering understandable as easily as possible, rather than to expound the analysis rigorously or to describe the applications in detail. The present version of my lecture notes has a similar aim. It is intended for senior students and for graduate students, phyicists, chemists and engineers as well as mathematicians. The book will also help specialists in these and other subjects who wish to become acquainted with the theory of solitons, but does not go as far as the rapidly advancing frontier of research. The fundamentals are introduced from the point of view of a course of advanced calculus or the mathematical methods of physics. Some knowledge of the elements of the theories of linear waves, partial differential equations, Fourier integrals, the calculus of variations, Sturm-Liouville theory and the hypergeometric function, but little more, is assumed. Also some familiarity with the elements of the theories of water waves, continuous groups, elliptic functions, one-dimensional wave mechanics and Hilbert spaces will be useful, but is not essential. References are given to help those readers who have not learnt these topics. The diverse applications of the theory of solitons are only mentioned briefly in the main text and in the problems.

Simplicity and concrete applications are emphasized in order to make the material easily assimilable. The more difficult sections, paragraphs and problems are preceded by asterisks; it is suggested that they are omitted on a first reading of the book. Further reading is recommended to cover results which are only quoted here and to offer more detailed treatments. The equations are numbered in each section separately. An equation within the same section is referred to simply by its number; in a different section by its section and equation numbers; and in a different chapter by its chapter, section and equation numbers. The sections and figures are numbered in each chapter separately and are referred to similarly.

I am grateful to Miss Sarah Trickett for computing the solutions of the Korteweg-de Vries equation used to draft Figures 4.6, 4.7 and 4.8; to Drs A. Davey, D.H. Griffel, R.S. Johnson, I.M. Moroz and A.R. Paterson for criticisms of an earlier draft of this book; to Mrs. Tina Harrison for drawing the figures; and to Mrs Nancy Thorp for typing the text — its quality is immediately evident to the reader although my last-minute changes may not be.

P.G. Drazin
Bristol
February 1983

CHAPTER 1 THE KORTEWEG-DE VRIES EQUATION

1 *The discovery of solitary waves*

This book is an introduction to the theory of solitons and to the applications of the theory. Solitons are a special kind of localized wave, an essentially nonlinear kind. We shall define them at the end of this chapter, describing their discovery by Zabusky & Kruskal (1965). A *solitary wave* is the first and most celebrated example of a soliton to have been discovered, although more than 150 years elapsed after the discovery before a solitary wave was recognized as an example of a soliton. To lead to the definition of a soliton, it is helpful to study solitary waves on shallow water. We shall describe briefly in this section the properties of these waves, and then revise the elements of the theory of linear and nonlinear waves in order to build a foundation of the theory of solitons. Let us begin at the beginning, and relate a little history.

The solitary wave, or great wave of translation, was first observed on the Edinburgh to Glasgow canal in 1834 by J. Scott Russell. Russell reported his discovery to the British Association in 1844 as follows:

> I believe I shall best introduce this phænomenon by describing the circumstances of my own first acquaintance with it. I was observing the motion of a boat which was rapidly drawn along a narrow channel by a pair of horses, when the boat suddenly stopped — not so the mass of water in the channel which it had put in motion; it accumulated round the prow of the vessel in a state of violent agitation, then suddenly leaving it behind, rolled forward with great velocity, assuming the form of a large solitary elevation, a rounded, smooth and well-defined heap of water, which continued its course along the channel apparently without change of form or diminution of speed. I followed it on horseback, and overtook it still rolling on at a rate of some eight or nine miles an hour, preserving its original figure some thirty feet long and a foot to a foot and a half in height. Its height

> gradually diminished, and after a chase of one or two miles I lost it in the windings of the channel.

Russell also did some laboratory experiments, generating solitary waves by dropping a weight at one end of a water channel (see Fig. 1). He deduced empirically that the volume of water in the wave is equal to the volume displaced by the weight and that the steady velocity c of the wave is given by

$$c^2 = g(h + a), \tag{1}$$

where the amplitude a of the wave and the height h of the undisturbed water are as defined in Fig. 2. Note that a taller solitary wave travels faster than a smaller one. Russell also made many other observations and experiments on solitary waves. In particular, he tried to generate waves of depression by raising the weight from the bottom of the channel initially. He found, however, that an initial depression becomes a train of oscillatory waves whose lengths increase and amplitudes decrease with time (see Fig. 3).

Boussinesq (1871) and Rayleigh (1876) independently assumed that a solitary wave has a length much greater than the depth of the water

Fig. 1. Russell's solitary wave: a diagram of its development. (a) The start. (b) Later. (After Russell 1844.)

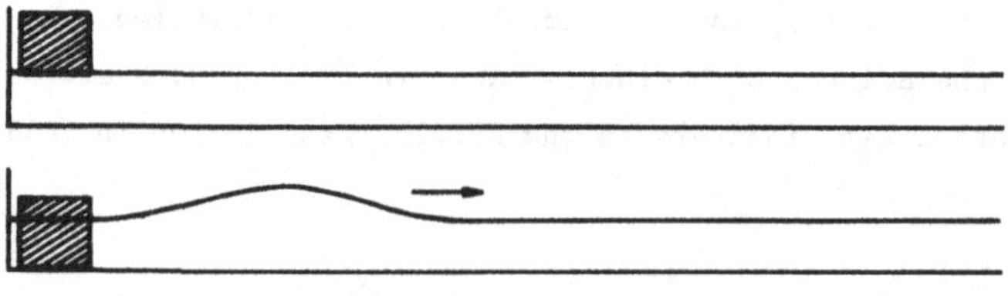

Fig. 2. The configuration and parameters for description of a solitary wave.

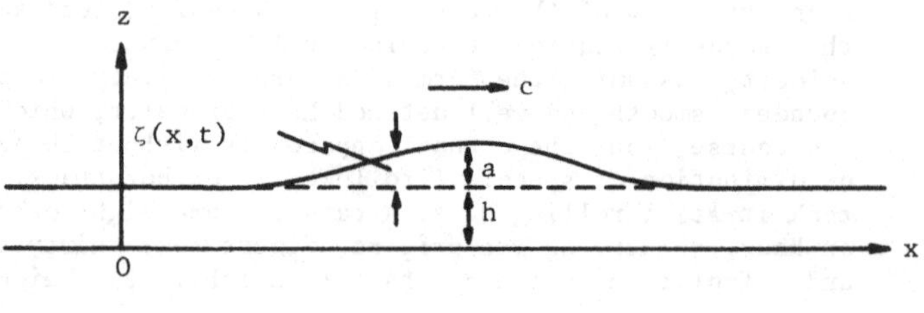

and thereby deduced Russell's empirical formula for c from the equations of motion of an inviscid incompressible fluid. They further showed essentially that the wave height above the mean level h is given by

$$\zeta(x,t) = a\,\mathrm{sech}^2\{(x - ct)/b\}, \tag{2}$$

where $b^2 = 4h^2(h + a)/3a$ for any positive amplitude a.

In 1895 Korteweg and de Vries developed this theory, and found an equation governing the two-dimensional motion of weakly nonlinear long waves:

$$\frac{\partial\zeta}{\partial t} = \frac{3}{2}\sqrt{\frac{g}{h}}\left(\zeta\frac{\partial\zeta}{\partial x} + \frac{2}{3}\alpha\frac{\partial\zeta}{\partial x} + \frac{1}{3}\sigma\frac{\partial^3\zeta}{\partial x^3}\right), \tag{3}$$

where α is a small but otherwise arbitrary constant, $\sigma = \frac{1}{3}h^3 - Th/g\rho$, and T is the surface tension of the liquid of density ρ. This is essentially the original form of the *Korteweg-de Vries equation*; we shall call it the *KdV equation*. Note that in the approximations used to derive this equation one considers long waves propagating in the direction of increasing x. A similar equation, with $-\partial\zeta/\partial t$ instead of $\partial\zeta/\partial t$, may be applied to waves propagating in the opposite direction.

Fig. 3. Russell's observations of oscillating waves: successive stages of their development. (After Russell 1844.)

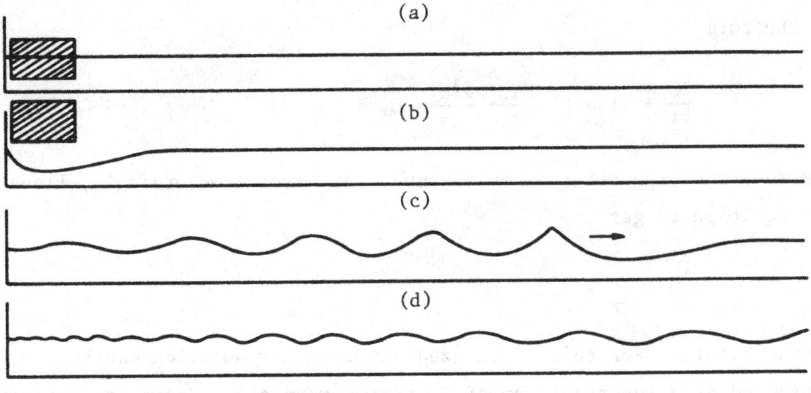

2 *Fundamental ideas*

We shall dwell on the mathematical ideas, rather than the applications to water waves, in this book. First note that by translations and magnifications of the dependent and independent variables,

$$u = k_1\zeta + k_o, \quad X = k_3x + k_2, \quad T = k_4t + k_5, \tag{1}$$

we can write the KdV equation in many equivalent forms by choice of the constants k_o to k_5:

$$\frac{\partial u}{\partial T} = (1 + u)\frac{\partial u}{\partial X} + \frac{\partial^3 u}{\partial X^3}, \tag{2}$$

$$\frac{\partial u}{\partial T} + (1 + u)\frac{\partial u}{\partial X} + \frac{\partial^3 u}{\partial X^3} = 0, \tag{3}$$

$$\frac{\partial u}{\partial T} - 6u\frac{\partial u}{\partial X} + \frac{\partial^3 u}{\partial X^3} = 0, \tag{4}$$

$$\frac{\partial u}{\partial T} + 6u\frac{\partial u}{\partial X} + \frac{\partial^3 u}{\partial X^3} = 0, \text{ etc.} \tag{5}$$

(These transformations are examples of *Lie groups* or *continuous groups*, which are the subject of an extensive theory and which have many applications, notably in physics. For further reading, the book by Bluman & Cole (1974) is recommended.) We shall usually use equation (4) as the standard form.

To understand how solitons may persist, take the KdV equation in the form

$$\frac{\partial u}{\partial t} + (1 + u)\frac{\partial u}{\partial x} + \frac{\partial^3 u}{\partial x^3} = 0, \tag{6}$$

and seek the properties of small-amplitude waves. Accordingly, linearize the equation to get

$$\frac{\partial u}{\partial t} + \frac{\partial u}{\partial x} + \frac{\partial^3 u}{\partial x^3} = 0, \tag{7}$$

approximately. For this linearized equation any solution can be represented as a superposition of Fourier components. So we use the method of normal modes, with independent components $u \propto e^{i(kx-\omega t)}$. It follows that

$$\omega = k - k^3. \tag{8}$$

This is the *dispersion relation* which gives the frequency ω as a function of the wavenumber k. From it we deduce the *phase velocity*,

$$c = \frac{\omega}{k} = 1 - k^2, \tag{9}$$

which gives the velocity of the wave fronts of the sinusoidal mode. We also deduce the *group velocity*,

$$c_g = \frac{d\omega}{dk} = 1 - 3k^2, \tag{10}$$

which gives the velocity of a wave packet, i.e. a group of waves with nearly the same length $2\pi/k$. Note that $c_g \leqslant c \leqslant 1$, and $c = c_g = 1$ for long waves (i.e. for $k = 0$). Also a short wave has a negative phase velocity c.

Packets of waves of nearly the same length propagate with the group velocity, individual components moving through the packet with their phase velocity. It can in general be shown that the energy of a wave disturbance is propagated at the group velocity, not the phase velocity. Long-wave components of a general solution travel faster than the short-wave components, and thereby the components disperse. Thus the linear theory predicts the dispersal of any disturbance other than a purely sinusoidal one. Looking back to the equation, you can see that the dispersion comes from the term in k^3 in the expression for ω and thence from the term $\partial^3u/\partial x^3$ in the KdV equation.

For further reading on group velocity, Lighthill's (1978, §3.6) book is recommended.

In contrast to dispersion, nonlinearity leads to the concentration of a disturbance. To see this, neglect the term $\partial^3u/\partial x^3$ in the KdV equation above and retain the nonlinear term. Then we have

$$\frac{\partial u}{\partial t} + (1 + u)\frac{\partial u}{\partial x} = 0. \tag{11}$$

The method of characteristics may be used to show that this equation has the elementary solution

$$u = f\{x - (1 + u)t\} \tag{12}$$

for any differentiable function f. (You may also verify that this satisfies equation (11).) This shows that disturbances travel at the characteristic velocity $1 + u$. Thus the 'higher' parts of the solution travel faster than the 'lower'. This 'catching up' tends to steepen a disturbance until it 'breaks' and a discontinuity or *shock wave* forms (see Fig. 4).

For further reading on wave breaking, the books by Landau & Lifshitz (1959, §94) and Whitham (1974, §2.1) are recommended.

We anticipate that for a solitary wave the dispersive effects of the term $\partial^3u/\partial x^3$ and the concentrating effects of the term $u\partial u/\partial x$ are just in balance. We shall examine the details of this balance in the next few chapters. A similar balance occurs for a large number of solutions of nonlinear equations, a few of which will be demonstrated in the text and problems of this book.

In future, we shall usually denote partial differentiation by a subscript, so that $u_t = \partial u/\partial t$, $u_x = \partial u/\partial x$, $u_{xxx} = \partial^3u/\partial x^3$ etc.

Fig. 4. Sketch of the nonlinear steepening of a wave as it develops. (a) t = 0. (b) Later.

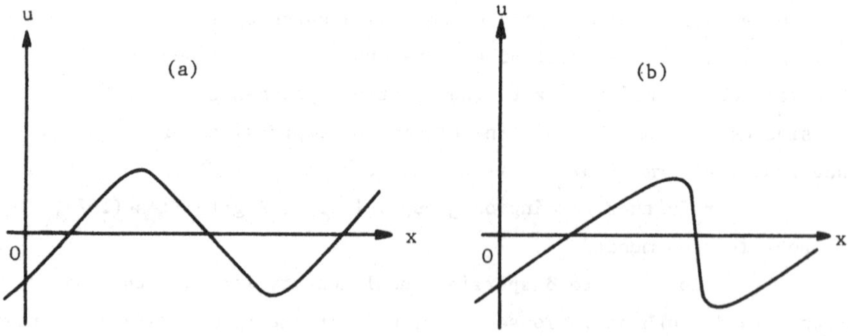

3 *The discovery of soliton interactions*

Examining the Fermi-Pasta-Ulam model of phonons in an anharmonic lattice, Zabusky & Kruskal (1965) were led to work on the KdV equation. They considered the following initial-value problem in a periodic domain:

$$\frac{\partial u}{\partial t} + u\frac{\partial u}{\partial x} + \delta\frac{\partial^3 u}{\partial x^3} = 0 \tag{1}$$

i.e. $u_t + uu_x + \delta u_{xxx} = 0,$

where

$$u(2,t) = u(0,t), \quad u_x(2,t) = u_x(0,t), \quad u_{xx}(2,t) = u_{xx}(0,t) \quad \text{for all } t, \tag{2}$$

and $u(x,0) = \cos\pi x \quad \text{for } 0 \leqslant x \leqslant 2.$ (3)

Fig. 5. Solutions of the KdV equation $u_t + uu_x + \delta u_{xxx} = 0$ with $\delta = 0.022$ and $u(x,0) = \cos\pi x$ for $0 \leqslant x \leqslant 2$. (After Zabusky & Kruskal 1965.) (a) The dotted curve gives u at $t = 0$. (b) The broken curve gives u at $t = 1/\pi$. Note that a 'shock wave' has nearly formed at $x = 0.5$; this is because the initial solution is not steep enough for dispersion to be significant (i.e. because $|\delta u_{xxx}| \ll |uu_x|$ mostly) and therefore the terms $u_t + uu_x$ have been approximately zero up till this time. (c) The continuous curve gives u at $t = 3.6/\pi$. Note the formation of eight more-or-less distinct solitons, whose crests lie close to a straight line (extended with the period 2).

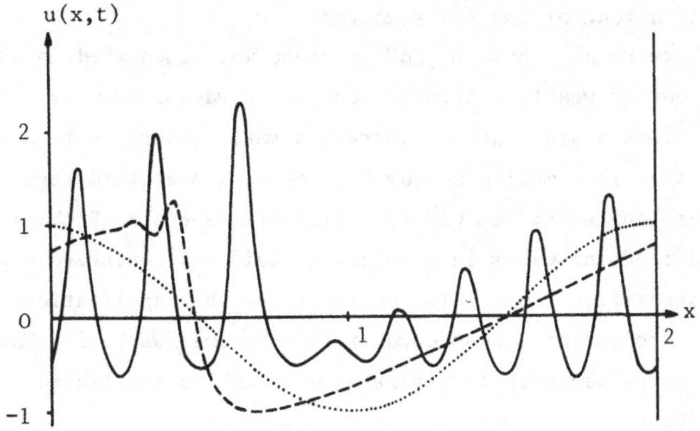

The periodic boundary conditions suit numerical integration of the system. Putting $\delta = 0.022$, Zabusky & Kruskal computed u for $t > 0$. They found that the solution breaks up into a train of eight solitary waves (see Fig. 5), each like a sech-squared solution, that these waves move through one another as the faster ones catch up the slower ones, and that finally the initial state (or something very close to it) recurs. This remarkable numerical discovery, that strongly nonlinear waves can interact and carry on thereafter almost as if they had never interacted, led to an intense study of the analytic and numerical properties of many kinds of *solitons*. This intense study continues still.

A 'soliton' is not precisely defined, but is used to describe any solution of a nonlinear equation or system which (i) represents a wave of permanent form; (ii) is localized, decaying or becoming constant at infinity; and (iii) may interact strongly with other solitons so that after the interaction it retains its form, almost as if the principle of superposition were valid. The word 'soliton' was coined by Zabusky & Kruskal (1965) after 'photon', 'proton', etc. to emphasize that a soliton is a localized entity which may keep its identity after an interaction. (The Greek word 'on' means 'solitary'.) The word may also symbolize the hope that the properties of elementary particles will be deduced by calculation of soliton solutions of some nonlinear field theory.

A solitary wave may be defined more generally than as a sech-squared solution of the KdV equation. We take it to be any solution of a nonlinear system which represents a hump-shaped wave of permanent form, whether it is a soliton or not.

4 *Applications of the KdV equation*

We have related how the KdV equation was discovered in 1895 to model the behaviour of weakly nonlinear long water waves. Benney (1966) recognized that this approximation, whereby a small quadratic term representing convection in a moving medium balances a linear term representing dispersion of long waves, is widely applicable (see also Problem 1.8). He applied it to inertial waves in a rotating fluid and to internal gravity waves in a stratified fluid. Two of the many other applications of the KdV equation are to ion-acoustic waves in a plasma (Washimi & Taniuti 1966) and to pressure waves in a misture of gas bubbles and liquid (Wijngaarden 1968).

Problems

1.1 *Motion pictures of soliton interactions.* See the animated computer films of solitons by Zabusky, Kruskal & Deem (F1965) and Eilbeck (F1981). These films are listed in the motion picture index.

1.2 *The 'tail' of a solitary wave.* Verify that $u(x,t) = -2\kappa^2\text{sech}^2\{\kappa(x - 4\kappa^2 t)\}$ satisfies the KdV equation in the form

$$\frac{\partial u}{\partial t} - 6u\frac{\partial u}{\partial x} + \frac{\partial^3 u}{\partial x^3} = 0,$$

i.e. $$u_t - 6uu_x + u_{xxx} = 0.$$

Show that $u(x,t) = e^{2\kappa(x-ct)}$ is a solution of the linearized KdV equation,

$$u_t + u_{xxx} = 0,$$

if $c = 4\kappa^2$. How is the latter solution related to the former as $x \to -\infty$?

1.3 *The formation of a 'shock wave'.* Verify that if $u = \cos\{\pi(x - ut)\}$ then

$$u_t + uu_x = 0 \qquad \text{for } 0 \leqslant x < 2,$$

$u(x + 2,t) = u(x,t)$ for $t > 0$, and $u(x,0) = \cos\pi x$. Where (i.e. at what stations x) might u_x approach infinity? Show that u first ceases to be single valued when $t = 1/\pi$. How might one seek to continue the solution for $t > 1/\pi$?

1.4 *The initial-value problem for the linearized KdV equation.* If

$$u_t + u_{xxx} = 0,$$

$u(x,0) = g(x)$, and $u, u_x, u_{xx} \to 0$ as $x \to \pm\infty$, show that

$$u(x,t) = \frac{1}{2\pi}\int_{-\infty}^{\infty} \exp\{ik(x + k^2 t)\}\int_{-\infty}^{\infty} g(y)e^{-iky}dy\, dk$$

$$= (3t)^{-1/3}\int_{-\infty}^{\infty} g(y)\frac{1}{2\pi}\int_{-\infty}^{\infty} \exp\left[i\left\{\frac{\alpha(x-y)}{(3t)^{1/3}} + \frac{1}{3}\alpha^3\right\}\right]d\alpha dy.$$

Deduce that

$$u(x,t) \sim \int_{-\infty}^{\infty} g(y)dy(3t)^{-1/3}\mathrm{Ai}(X) - \int_{-\infty}^{\infty} yg(y)dy(3t)^{-2/3}\mathrm{Ai}'(X) + \ldots$$

as $t \to \infty$ for fixed $X = x/(3t)^{1/3}$, where Ai is the Airy function. Show that the first term of the above asymptotic expansion represents a steeply rising wave front where $x \approx t^{1/3}$ and a slowly decaying wave train where $x \lesssim -t^{1/3}$.

[You are given that $\mathrm{Ai}(X) = \frac{1}{2\pi}\int_{-\infty}^{\infty} \exp\left\{i(\alpha X + \frac{1}{3}\alpha^3)\right\}d\alpha$.]

1.5 *The essential nonlinearity of a solitary wave.* Why is it that a solitary wave of infinitesimal amplitude persists whereas all localized solutions of the linearized KdV equation disperse and decay in time? Discuss.

1.6 *Similarity solutions of the KdV equation.* Show that the KdV equation,

$$u_t - 6uu_x + u_{xxx} = 0,$$

is invariant under the one-parameter continuous group of transformations $x \to kx$, $t \to k^3t$ and $u \to k^{-2}u$. Deduce that $t^{2/3}u$ and $x/t^{1/3}$ are also invariant under this group.

Assuming that there exists a solution of the form

$$u(x,t) = -(3t)^{-2/3}U(X), \text{ where } X = x/(3t)^{1/3},$$

show that

$$\frac{d^3U}{dX^3} + (U - X)\frac{dU}{dX} - 2U = 0.$$

[Miura (1976, p.437).]

1.7 *The RLW equation.* The *regularized long-wave equation, RLW equation* or *Peregrine equation,* namely

$$\zeta_t + \sqrt{\frac{g}{h}}(h + \frac{3}{2}\zeta)\zeta_x - \frac{1}{6}h^2\zeta_{xxt} = 0,$$

describes long waves on water of depth h (much as the KdV equation does). Linearize it for small wave height ζ, assume that $\zeta \propto e^{i(kx-\omega t)}$, and deduce the dispersion relation. Evaluate the phase and group velocities of these waves. Compare your results with those for infinitesimal water waves of arbitrary length, for which the dispersion relation is given to be $\omega^2 = gk\tanh(kh)$, noting the similarities for long waves.

[Peregrine (1966); Benjamin, Bona & Mahony (1972).]

*1.8 *Derivation of equations of weakly nonlinear long waves.* Suppose that the phase velocity of some linear wave of length $2\pi/k$ is $c(k)$. Then weakly nonlinear long waves are often describable by an equation of the form

$$u_t + Auu_x + \int_{-\infty}^{\infty} K(x-\xi)u_\xi(\xi,t)d\xi = 0,$$

where the kernel K is determined by linear theory as the Fourier transform of c, namely

$$K(x) = \frac{1}{2\pi}\int_{-\infty}^{\infty} c(k)e^{ikx}dk,$$

and A is some constant.

Use this equation and Fourier analysis to justify the KdV equation for weakly nonlinear long water waves.

Also take $c(k) = c_0(1 - \gamma|k|)$ and deduce the *DABO equation,* or *Benjamin-Ono equation,* namely

$$u_t + (c_0 + Au)u_x + \frac{\gamma c_0}{\pi} P\int_{-\infty}^{\infty} \frac{u_{\xi\xi}(\xi,t)}{\xi - x} d\xi = 0,$$

where P denotes the Cauchy principal value of the integral.

[Benjamin (1967, pp.562-3); Whitham (1967, p.22); Whitham (1974, §13.14); Titchmarsh (1948, p.120).]

1.9 *The modulation of a wave packet.* A real linear partial differential equation is given in the operational form,

$$\phi_t + if(-i\partial/\partial x)\phi = 0,$$

where f is an odd function. Show that wave solutions $\phi(x,t) = \mathrm{Re}\{e^{i(kx-\omega t)}\}$ have the dispersion relation $\omega = f(k)$.

Consider a wave packet described by the solution $\phi(x,t) = \mathrm{Re}\{A(x,t)\Phi(x,t)\}$, where the amplitude A is a slowly varying complex function, and the 'carrier wave' $\Phi = e^{i(Kx-\Omega t)}$ has frequency $\Omega = f(K)$ for a given wavenumber $K \neq 0$. Deduce that

$$i\{A_t + f'(K)A_x\} + \tfrac{1}{2}f''(K)A_{xx} = 0$$

approximately, if f has a Taylor series at K.

You are given further that the addition of some nonlinear terms to the equation for ϕ results, at leading order for small amplitudes A, in the addition of a term proportional to $|A|^2A$ to the modulation equation for A, so weakly nonlinear wave packets are governed by the equation,

$$i\{A_t + f'(K)A_x\} + \tfrac{1}{2}f''(K)A_{xx} + \ell|A|^2A = 0.$$

Remove the term in A_x by a Galilean transformation (noting that $f'(K)$ is the group velocity of the linear wave packet) and rescale x and t to show that this equation is equivalent to the *cubic Schrödinger equation*, namely

$$iA_T + A_{XX} + \nu|A|^2A = 0.$$

Now show that this equation has solution $A(X,T) = a\,e^{i(\kappa X-nT)}$ for all real a and κ provided that $n = \kappa^2 - \nu a^2$.

[Similar arguments may be applied to many more-general nonlinear partial differential systems to show in detail that weakly nonlinear wave packets are governed by the cubic Schrödinger equation. Notable examples are in the theories of nonlinear optics (Kelley 1965) and water waves (Davey 1972, Hasimoto & Ono 1972).]

CHAPTER 2 CNOIDAL WAVES

1 *Wave solutions of the KdV equation*

It is convenient to use the KdV equation in the standard form,

$$\frac{\partial u}{\partial t} - 6\,u\,\frac{\partial u}{\partial x} + \frac{\partial^3 u}{\partial x^3} = 0. \tag{1}$$

To solve it, first seek waves of permanent shape and size by trying solutions such that

$$u(x,t) = f(X) \quad \text{and} \quad X = x - ct, \tag{2}$$

for some function f and constant wave velocity c. We determine f and c by substitution of the assumed form of solution into the KdV equation. This gives

$$-\,cf' - 6ff' + f''' = 0$$

and therefore

$$\frac{d\tfrac{1}{2}f'^2}{df} = f'' = 3f^3 + cf + A$$

for some constant A of integration. Similarly,

$$\tfrac{1}{2}f'^2 = f^3 + \tfrac{1}{2}cf^2 + Af + B. \tag{3}$$

2 *Solitary waves*

If we seek a solitary wave, we may add the boundary conditions that $f, f', f'' \to 0$ as $X \to \pm\infty$. Therefore $A, B = 0$ and

$$\tfrac{1}{2}f'^2 = f^2(f + \tfrac{1}{2}c). \tag{1}$$

Therefore

$$X = \int \frac{df}{f'} = \int \frac{df}{f\sqrt{2f + c}} \ . \tag{2}$$

The substitution $f = -\tfrac{1}{2}c\ \mathrm{sech}^2\theta$ then gives the solution

$$f(X) = -\tfrac{1}{2}c\ \mathrm{sech}^2\{\tfrac{1}{2}\sqrt{c}(X - X_0)\} \tag{3}$$

for any constants $c \geqslant 0$ and X_0. Note that this soliton is a wave of depression, i.e. $u < 0$ for all X. It would have been a wave of elevation, like the one Scott Russell saw, if we had begun with the KdV equation in the form

$$\frac{\partial u}{\partial t} + 6u\frac{\partial u}{\partial x} + \frac{\partial^3 u}{\partial x^3} = 0.$$

The amplitude $\tfrac{1}{2}c$ of the soliton is proportional to its velocity. The width is inversely proportional to $\sqrt{c}$. Thus deep waves are narrow and fast.

It is interesting to note that such solitary waves, although essentially nonlinear phenomena, exist however small their amplitudes are.

3 *General waves of permanent form*

The *qualitative* behaviour of the solutions f for general values of the constants A and B can be found by elementary analysis, although numerical work is needed to find the quantitative behaviour.

The important point to remember in solving

$$\tfrac{1}{2}f'^2 = f^3 + \tfrac{1}{2}cf^2 + Af + B, \equiv F(f), \text{ say,} \tag{1}$$

is that we seek real bounded solutions $f(X)$. Therefore $f'^2 \geqslant 0$ and f increases monotonically until f' vanishes. Therefore the zeros of the cubic F on the right-hand side of equation (1) are important.

The qualitative character of the zeros of F according to the values of c, A and B must be essentially one of six types, indicated by the graphs of F in Fig. 1. Remember that $f'^2 \geqslant 0$, so that a real solution f is permitted only in an interval which is shaded.

Next find the behaviour of f near a zero, say f_1, of f'. First suppose that f_1 is a simple zero of F. Then

$$f'^2 = 2F'(f_1)(f - f_1) + O\{(f - f_1)^2\} \quad \text{as} \quad f \to f_1. \tag{2}$$

Therefore

$$f = f_1 + \tfrac{1}{2}F'(f_1)(X - X_1)^2 + O\{(X - X_1)^3\} \quad \text{as} \quad X \to X_1, \tag{3}$$

where X_1 is the value of X where $f = f_1$, i.e. $f(X_1) = f_1$. It follows that f has a simple minimum or maximum f_1 at X_1, according as $F'(f_1)$ is positive or negative respectively.

Fig. 1. Sketches of the graphs of F(f) for the different cases. The 'admissible' intervals of f where $f'^2 = 2F > 0$ are dotted. A bounded solution may exist on such a bounded interval.

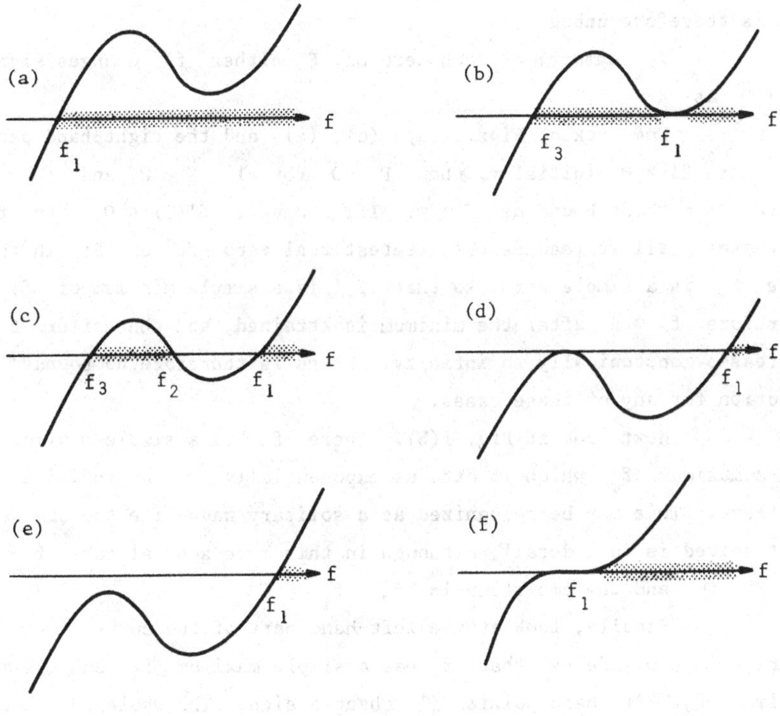

Similarly, if f_1 is a double zero of F, then

$$f'^2 = F''(f_1)(f - f_1)^2 + O\{(f - f_1)^3\} \quad \text{as} \quad f \to f_1. \tag{4}$$

This can occur for real f' only if $F''(f_1) > 0$, as illustrated in Fig. 1(b), and gives

$$f - f_1 \sim \text{constant} \times \exp\{\pm\sqrt{F''(f_1)}X\} \quad \text{as} \quad X \to \mp\infty \tag{5}$$

respectively, in order that f is bounded. Thus the solution approaches its limit f_1, the double zero of F, in an infinite distance in X.

There is only one possibility of a triple zero of F, namely $f_1 = -\frac{1}{6}c$. This is illustrated in Fig. 1(f). It can easily be shown that in this case the exact solution is of the form

$$f(X) = -\frac{1}{6}c + 2/(X - X_0)^2,$$

and is therefore unbounded.

In summary, at each zero of F either f' changes sign or $f' \to 0$ as $X \to \pm\infty$.

Now look at Figs. 1(a), (d), (e) and the right-hand part of (c). If $f' > 0$ initially, then $F > 0$ for all $X > 0$, and f increases without bound as $X \to \infty$. If, however, $f'(0) < 0$, then f decreases until it reaches the greatest real zero f_1 of F; in this case f_1 is a simple zero, so that f_1 is a simple minimum of f; therefore $f' > 0$ after the minimum is attained, and thereafter f increases monotonically to infinity. There is therefore no bounded solution for any of these cases.

Next look at Fig. 1(b). There f has a simple minimum f_3 and a maximum f_1 which it attains exponentially over an infinite distance. This may be recognized as a solitary wave like the one we have just solved in full detail, although in this more general case $f \to f_3$ as $X \to \pm\infty$ and the amplitude is $f_3 - f_1$.

Finally, look at the left-hand part of the curve of Fig. 1(c) where $f'^2 \geqslant 0$. We see that f has a simple maximum f_2 and a simple minimum f_3. At these points f' changes sign. The whole solution $f(X)$

is specified by the value of f and the sign of f at any point X_o because the equation is of first order. Therefore the solution is periodic. Its qualitative character can be seen to be like that in Fig.2. Further the period in X is

$$2\int_{f_3}^{f_2} \frac{df}{f'} = 2\int_{f_3}^{f_2} \frac{df}{\sqrt{2F(f)}}\ , \tag{6}$$

and the solution is given implicitly by

$$X = X_3 + \int_{f_3}^{f} \frac{df}{\pm\sqrt{2F(f)}} \tag{7}$$

on taking $\pm$ according to whether $f' \gtrless 0$ respectively, where $f(X_3) = f_3$.

These are called *cnoidal waves* because they can be described in simple terms of the elliptic function cn. They were originally found by Korteweg and de Vries themselves in 1895.

It can in fact be shown that both solitary waves (Benjamin 1972) and cnoidal waves (Drazin 1977) are stable solutions of the KdV equation.

Fig. 2. A cnoidal wave. (Only part of the infinite train is sketched.)

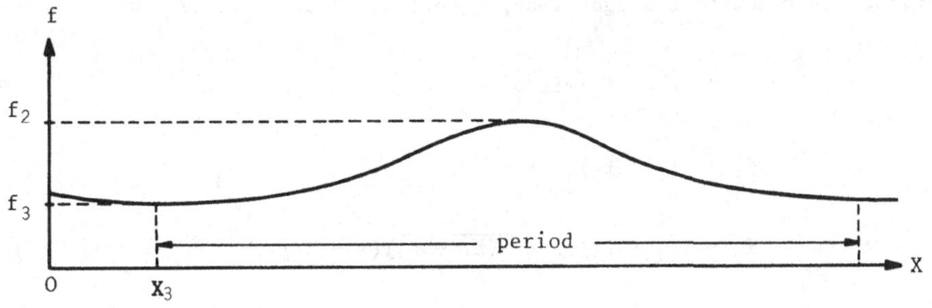

*4 *Description of waves in terms of elliptic functions*

We shall now digress to discuss Jacobian elliptic functions. We shall only mention a few properties relevant to cnoidal waves, and use the notation of Milne-Thomson. This notation and many formulae are given by Abramowitz & Stegun (1964), who also give many references for further reading. The qualitative description of cnoidal waves in §3 is, however, sufficient for many purposes.

We can express the solution as

$$X = X_3 + \int_{f_3}^{f} \frac{df}{\pm\sqrt{2F(f)}}$$

$$= X_3 + \int_{f_3}^{f} \frac{df}{\pm\sqrt{2(f - f_1)(f - f_2)(f - f_3)}} \tag{1}$$

in the case of Fig. 1(c), where $f_3 < f_2 < f_1$ are the three distinct real zeros of F. The integral may be identified as an *elliptic integral of the first kind* (Abramowitz & Stegun 1964, §§17.4.62). Accordingly define v and ϕ so that

$$v = \sqrt{\tfrac{1}{2}(f_1 - f_3)}(X - X_3) \tag{2}$$

and

$$\mathrm{cn}\, v = \cos\phi. \tag{3}$$

Therefore (Abramowitz & Stegun 1964, §§16.1.5, 17.2.6, 17.2.7)

$$f = f_3 + (f_2 - f_3)\sin^2\phi$$

$$= f_3 + (f_2 - f_3)(1 - \mathrm{cn}^2 v)$$

$$= f_2 - (f_2 - f_3)\mathrm{cn}^2(\sqrt{\tfrac{1}{2}(f_1 - f_3)}(X - X_3)\,|\,m), \tag{4}$$

where the parameter of the Jacobian elliptic function cn is $m = (f_2 - f_3)/(f_1 - f_3)$. This expresses the solution in explicit terms of the tabulated function cn.

We have needed to introduce the elliptic function so let us

digress in this paragraph to explain the few of its properties we use. The elliptic function $\mathrm{cn}(v|m)$ is defined by

$$v = \int_0^{\phi} (1 - m \sin^2\theta)^{-\frac{1}{2}} d\theta \quad \text{and} \quad \mathrm{cn}\, v = \cos\phi \tag{5}$$

(Abramowitz & Stegun 1964, §§16.1.3, 16.1.5). Note that if $m = 0$ then $v = \phi$ and $\mathrm{cn} = \cos$. More generally, if $0 < m < 1$ then cn is an analogue of the cosine defined by an integral like that for the inverse cosine. The graph of cn for $m = \frac{1}{2}$ is shown in Fig. 16.1 of Abramowitz & Stegun. If $m = 1$, then

$$v = \int_0^{\phi} \frac{d\theta}{\cos\theta} \;;$$

substitute w for θ, where $\cos\theta = \mathrm{sech}\, w$; therefore $-\sin\theta d\theta = -\sinh w\, \mathrm{sech}^2 w\, dw$; therefore

$$v = \int_0^{\mathrm{sech}^{-1}(\cos\phi)} \frac{\sinh w\, \mathrm{sech}^2 w\, dw}{\mathrm{sech}\, w \sqrt{1 - \mathrm{sech}^2 w}}$$

$$= \int_0^{\mathrm{sech}^{-1}(\mathrm{cn} v)} dw;$$

we conclude that if $m = 1$ then

$$\mathrm{cn} v = \mathrm{sech} v.$$

The *complete elliptic integral of the first kind* is defined (Abramowitz & Stegun 1964, §16.1.1) by

$$K(m) = \int_0^{\frac{1}{2}\pi} (1 - m \sin^2\theta)^{-\frac{1}{2}} d\theta, \tag{6}$$

where m is called the parameter (not the modulus). Note that

$$K(0) = \tfrac{1}{2}\pi, \quad K(m)\uparrow \quad \text{and} \quad K(1) = \infty.$$

The period of $\cos\phi$ is 2π and therefore the period of cn for $0 \lesssim m < 1$ is given by

$$4\int_0^{\frac{1}{2}\pi} \frac{d\theta}{(1 - m \sin^2\theta)^{\frac{1}{2}}} = 4K(m). \tag{7}$$

Of course, $K(1) = \infty$, which is what we expect to give the infinite 'period' of sech v.

We now end the digression and use the properties of the elliptic function cn to describe the cnoidal waves.

The parameters f_2 and f_3 determine the mean level and the amplitude of a cnoidal wave. The parameter X_3 determines the phase. Note that just as the period of $\cos^2\phi$ is π, and not 2π, so the period of f is

$$2K(m)\sqrt{\frac{2}{f_1 - f_3}}, = \frac{2\pi}{k} \quad \text{say.} \tag{8}$$

Thus we let $2\pi/k$ denote the wavlength of a nonlinear cnoidal wave.

We also define the frequency of the cnoidal wave as

$$\begin{aligned}\omega &= kc \\ &= -2k(f_1 + f_2 + f_3),\end{aligned} \tag{9}$$

on use of equation (3.1). Therefore

$$\begin{aligned}u(x,t) = f(X) &= f\{(kx - \omega t)/k\} \\ &= g(kx - \omega t), \quad \text{say,}\end{aligned}$$

where g has period 2π in x. Therefore g(X) has a Fourier expansion in {cosnX, sin nX} for n = 0,1,2, ... We thus have a wave of finite amplitude whose frequency ω depends not only on the wavenumber k but also on the amplitude $\frac{1}{2}(f_2 - f_3)$ and the other parameters.

*5 *Infinitesimal waves*

We next discuss an important limiting case of cnoidal waves. Consider small values of m. Now

$$f_3 \to f_2, \; c \to -2(f_1 + 2f_2) \qquad \text{as} \quad m \to 0$$

and

$$u(x,t) \to f_2 - (f_2 - f_3)\cos^2\{\sqrt{\tfrac{1}{2}(f_1 - f_2)}(X - X_3)\} \qquad \text{as} \quad m \to 0$$

$$= f_2 - (f_2 - f_3)\tfrac{1}{2}[1 + \cos\{\sqrt{2(f_1 - f_2)}(X - X_3)\}]$$

$$\to f_2 - \tfrac{1}{2}(f_2 - f_3)\cos\{\sqrt{2(f_1 - f_2)}(X - X_3)\}$$

$$= f_2 - a\cos\{k(X - X_3)\},$$

say, where

$$a = \tfrac{1}{2}(f_2 - f_3) \quad \text{and} \quad k = \sqrt{2(f_1 - f_2)}.$$

Note that in this limit as $a \sim \frac{1}{2}(f_1 - f_2)m \to 0$ we have

$$\omega = kc = \sqrt{2(f_1 - f_2)}\{-2(f_1 + 2f_2)\}$$

$$= -6f_2k - k^3.$$

This is the solution for infinitesimal waves. To verify this directly by linearization of the KdV equation for small u (and therefore small a), note that

$$u_t - 6uu_x + u_{xxx}$$

$$= -a\omega\sin\{k(X - X_3)\} - 6f_2 \times ak\sin\{k(X - X_3)\}$$

$$- ak^3\sin\{k(X - X_3)\} + O(a^2) \qquad \text{as} \quad a \to 0$$

$$= O(a^2).$$

*6 *Solitary waves again*

Another important limiting case arises when m is close to one. We find that

$$f_2 \to f_1,\ c \to -2(2f_1 + f_3)$$

and

$$f \to f_1 - (f_1 - f_3)\mathrm{sech}^2\{\sqrt{\tfrac{1}{2}(f_1 - f_3)}(X - X_3)\} \qquad \text{as} \quad m \to 1.$$

This is essentially the Boussinesq-Rayleigh solution for the solitary wave.

Problems

2.1 *Solitons for a modified KdV equation.* Find all solutions of

$$u_t + 6u^2u_x + u_{xxx} = 0$$

such that $u = f(x - ct)$ and u and its derivatives vanish as $x \to \pm\infty$ for all t, where f is some function and c is some constant.

[Zabusky (1967) showed how this equation may model the oscillations of a lattice of particles connected by nonlinear springs, as the Fermi-Pasta-Ulam model does.]

2.2 *Another solution of the modified KdV equation.* Show that a solution of the equation

$$u_t + 6u^2u_x + u_{xxx} = 0$$

is given by

$$u(x,t) = V - 4V/\{4V^2(x - 6V^2t)^2 + 1\}$$

for any real constant V.

[Zabusky (1967, p.236).]

2.3 *A general modified KdV equation.* If

$$u_t + (n + 1)(n + 2)u^nu_x + u_{xxx} = 0$$

for $n = 1,2, \ldots$, show that the relative magnitudes of the coefficients may be changed at will by magnifications of the scales of u, x and t and the sign of the ratios of the nonlinear term to the linear ones is invariant under the transformation $u \to -u$ if and only if n is even.

Show that solitary waves are given by

$$u^n = \tfrac{1}{2}c\,\text{sech}^2\{\tfrac{1}{2}n\sqrt{c}(x - ct - x_0)\}$$

for all $c > 0$. Deduce that these are waves of elevation with $u > 0$ if n is odd. Deduce further that if the sign of the coefficient of the nonlinear term in the modified KdV equation is instead negative then there is a change of sign of u if n is odd but no solitary wave if n is even.

[Zabusky (1967, p.236).]

2.4 *Similarity solutions of two modified KdV equations.* Show that the equations

$$u_t \pm 6u^2u_x + u_{xxx} = 0$$

are invariant under the transformations $x \to kx$, $t \to k^3t$, $u \to u/k$.

Defining the invariant independent variable $X = x/t^{1/3}$, and trying the invariant solution $u(x,t) = t^{-1/3}f(X)$, deduce that

$$f'' - \frac{1}{3}Xf' \pm 2f^3 = 0$$

when f vanishes sufficiently rapidly at infinity.

[The solutions f are known as *second Painlevé transcendents* (Ince 1927, p.345).]

2.5 *A breather solution of a modified KdV equation.* Transform the modified KdV equation,

$$u_t + 6u^2u_x + u_{xxx} = 0,$$

into

$$\frac{\partial}{\partial x}(v_t + 2v_x^3 + v_{xxx}) = 0,$$

by putting $u = v_x$. Show that if v and its derivatives vanish as $x \to \pm\infty$ and $\phi = \tan\frac{1}{2}v$ then

$$(1 + \phi^2)(\phi_t + \phi_{xxx}) + 6\phi_x(\phi_x^2 - \phi\phi_{xx}) = 0.$$

Hence or otherwise verify that

$$u(x,t) = -2\frac{\partial}{\partial x}\tan^{-1}\left\{\frac{\ell\sin(kx + mt + a)}{k\cosh(\ell x + nt + b)}\right\}$$

satisfies the modified KdV equation for all real k, ℓ, a and b if

$$m = k(k^2 - 3\ell^2) \quad \text{and} \quad n = \ell(3k^2 - \ell^2).$$

Show that the profile of the solution is an oscillation moving within a moving envelope shaped like a hyperbolic secant.

[This solution is an example of an oscillatory pulse, called a *breather* or a *bion*, which may interact as a soliton with a solitary wave or another breather.]

2.6 *The nonlinear Klein-Gordon equation.* Seek solutions of the equation,

$$\frac{\partial^2 u}{c^2 \partial t^2} = \nabla^2 u - V'(u),$$

of the form $u(x,t) = f(x - mt)$, where c is a given constant and V a given differentiable function. Show that $\frac{1}{2}(m^2/c^2 - 1)f'^2 + V(f) = E$ for some constant E, and discuss the qualitative properties of V giving rise to the location of zeros of $E - V$ and thence to periodic wave solutions or solitary waves.

2.7 *Solitary waves in an elastic medium.* Nonlinear dispersive waves in an elastic medium are governed by the equation

$$u_{tt} = u_{xx} + u_x u_{xx} + u_{xxxx}.$$

Find all solitary waves of the form $u = f(x - ct)$ such that all derivatives of f vanish at infinity (but f itself need not).

[Toda (1967b); Grindlay & Opie (1977).]

2.8 *Two-dimensional KdV equation.* Two-dimensional propagation of waves in shallow water is governed by the *Kadomtsev-Petviashvili equation*,

$$(u_t - 6uu_x + u_{xxx})_x + 3u_{yy} = 0.$$

Show that the linearized equation admits solutions of the form $u \propto e^{kx+\ell y-\omega t}$ if

$$\omega = k^3 + 3\ell^2/k.$$

Show that a solitary wave of the nonlinear equation is given by

$$u(x,y,t) = -\tfrac{1}{2}k^2\mathrm{sech}^2\{\tfrac{1}{2}(kx + \ell y - \omega t)\}.$$

[Kadomtsev & Petviashvili (1970) discovered the equation when working on plasma physics; Freeman (1980).]

2.9 *The DABO equation.* Verify that a solution of the nonlinear integro-differential equation,

$$u_t + uu_x + \{\mathcal{H}(u)\}_{xx} = 0,$$

is given by

$$u(x,t) = ba^2/\{(x - ct)^2 + a^2\},$$

and find the relationship between the constants a, b and c. Here the *Hilbert transform* $\mathcal{H}$ is defined by

$$\mathcal{H}(u) = \pi^{-1}P\int_{-\infty}^{\infty}\frac{u(y,t)}{y - x}dy,$$

where P denotes the Cauchy principal part of the integral.

[You are given that $\mathcal{H}\{a/(x^2 + a^2)\} = -x/(x^2 + a^2)$ if $a > 0$. The DABO equation governs weakly nonlinear internal gravity waves in an unbounded incompressible inviscid fluid of variable density. Davis & Acrivos (1967); Benjamin (1967); Ono (1975); Titchmarsh (1948,p.121).]

2.10 *The cubic Schrödinger equation.* Consider the equation

$$iu_t + u_{xx} + |u|^2u = 0$$

and seek waves of permanent form by trying the solution

$$u = re^{i(\theta + nt)},$$

where r and θ are real functions of $x - ct$ only, for some real constants n and c. Hence show that

$$\theta' = \tfrac{1}{2}(c + A/s) \quad \text{and} \quad s'^2 = -2F(s),$$

where $s = r^2$ and $F(s) = s^3 - 2(n - \tfrac{1}{4}c^2)s^2 - Bs + \tfrac{1}{4}A^2$ for some real constants A and B of integration.

Discuss briefly the occurrence and properties of periodic solutions according to the nature of the zeros of the cubic F. In particular show that there exist solitary waves of the form

$$u(x,t) = ae^{i\{\frac{1}{2}c(x-ct)+nt\}}\operatorname{sech}\{a(x - ct)/\sqrt{2}\}$$

for all $a^2 = 2(n - \tfrac{1}{4}c^2) > 0$.

2.11 *Vibrations of a monatomic lattice.* A crystal may be modelled by a lattice of equal particles of mass m. The interaction of neighbouring particles is represented by nonlinear springs. It is given that it follows from Newton's second law of motion that

$$m\ddot{u}_n = 2f(u_n) - f(u_{n+1}) - f(u_{n-1})$$

for $n = 0, \pm 1, \pm 2, \ldots,$ where u_n is the displacement of the nth particle from its position of equilibrium and f is the force function of the springs.

Linearize this system for small u_n, and show that there are travelling waves of the form

$$u_n = \varepsilon \exp(i\theta_n), \ \theta_n = \omega t - pn$$

›rovided that

$$m\omega^2 = -4\sin^2(\tfrac{1}{2}p)f'(0) > 0.$$

For the *Toda lattice* or *Toda chain*, take the force function (u) = $-a(1 - e^{-bu})$ for some constants $a,b > 0$. To solve the system in his case, define s_n by $\dot{s}_n = f(u_n)$ and by the condition that $s_n = 0$ hen $u_n = 0$, and seek a solution of the form $s_n = g(\theta_n)$. Deduce that

$$\frac{m\omega^2}{b}\,\frac{g''(\theta_n)}{a + \omega g'(\theta_n)} = g(\theta_n + p) + g(\theta_n - p) - 2g(\theta_n).$$

Show that there exist solitary-wave solutions of this differential-difference equation having the form $s_n = m\omega b^{-1}\tanh\theta_n$, if $m\omega^2 = ab\sinh^2 p$.

[Kittel (1976, Chap.4); Toda (1967a,b). In fact it can be shown (i) by use of elliptic functions that there are periodic nonlinear waves of permanent form, analogous to cnoidal waves; (ii) that the solitary waves above interact like solitons; and (iii) the continuous limit of the discrete system is governed by a partial differential equation of the form seen in Problem 2.7.]

*2.12 *Fisher's equation.* Seek travelling-wave solutions of the nonlinear diffusion equation,

$$u_t = u_{xx} + \alpha^2 u(1 - u) \quad \text{for} \quad \alpha > 0,$$

by trying $u(x,t) = f(X)$, where $X = x - ct$. Investigate the solutions in the phase plane of f and $g = f'$, showing that

$$\frac{dg}{df} = -\frac{\alpha^2 f(1 - f) + cg}{g}.$$

Applying the theory of nonlinear ordinary differential equations, show that the equilibrium point $f = 1,\ g = 0$ is a saddle point and that the equilibrium point $f = 0,\ g = 0$ is an unstable node if $c < -2\alpha < 0$. Deduce that there exist travelling waves such that $u \to 0$ as $x \to -\infty$ and $u \to 1$ as $x \to \infty$ for all $c \leqslant -2\alpha$.

[Fisher (1937); Kolmogorov, Petrovsky & Piscounov (1937); Aris (1975, §8.10). In fact this wave is stable if $c = -2\alpha$, but the waves are not solitons because they do not pass through one another and keep their identities after the interaction. Fisher (1937) devised the equation for the selection of a gene in a species, but it or similar equations have since been used in the theory of combustion, of chemical kinetics, and of other subjects (cf. Aris (1975), Fife (1979).)]

2.13 *The Burgers equation.* Show that the nonlinear diffusion equation

$$u_t + uu_x = \nu u_{xx}$$

has a solution of the form

$$u(x,t) = c[1 - \tanh\{c(x - ct)/2\nu\}]$$

for all c.

[The solution is equivalent to one Taylor (1910) used to describe the structure of a weak shock wave in a real fluid. The equation was formulated by Burgers (1948) in an (entirely separate) attempt to model turbulent flow in a channel. The solutions do not interact like solitons (see Problem 4.2).]

*2.14 *Modons.* It is given that the motion of a thin layer of an incompressible inviscid fluid on a rapidly rotating sphere is governed approximately by a two-dimensional vorticity equation of the form

$$\zeta_t - \frac{1}{R^2}\psi_t + \beta\psi_x + \psi_x\zeta_y - \psi_y\zeta_x = 0,$$

where ψ is a stream function and $\zeta = \psi_{xx} + \psi_{yy}$ is the relative vorticity. A locally Cartesian frame fixed to the sphere is used, with the x-axis pointing eastwards and the y-axis northwards. The constant R is a certain length, called the radius of deformation, and β is the value of the northward derivative of the Coriolis parameter at the latitude where $y = 0$.

Seeking a wave of permanent form, show that if $\psi(x,y,t) = g(x - ct,y)$ for some function g and constant velocity c, then

$$g_{xx} + g_{yy} - g/R^2 + \beta y = G(g + cy)$$

for some differentiable function G of integration.

Show that if

$$g(r,\theta) = ac\sin\theta \times \begin{cases} \dfrac{q^2 J_1(kr/a)}{k^2 J_1(k)} - \left(1 + \dfrac{q^2}{k^2}\right)\dfrac{r}{a} & \text{for } r < a \\ -\dfrac{K_1(qr/a)}{K_1(q)} & \text{for } r > a \end{cases} ,$$

where $x = r\cos\theta$ and $y = r\sin\theta$ and a, k and q are positive constants, then

$$\zeta(r,\theta) = -a^{-1}cq^2\sin\theta \times \begin{cases} J_1(kr/a)/J_1(k) & \text{for } r < a \\ K_1(qr/a)/K_1(q) & \text{for } r > a \end{cases} .$$

Deduce that this gives a solution of the vorticity equation with

$$G(g + cy) = \begin{cases} \{\beta/c - (k^2 + q^2)/a^2\}(g + cy) & \text{for } r < a \\ \beta(g + cy)/c & \text{for } r > a \end{cases}$$

if $q = a\sqrt{R^{-2} + \beta/c}$.

Verify that for this solution ψ, ψ_r and ζ are continuous at $r = a$ if k is one of the (countable infinity of) roots of

$$\frac{K_2(q)}{qK_1(q)} = -\frac{J_2(k)}{kJ_1(k)} .$$

[These localized vortices are called *modons*. They may be important in meteorology and oceanography. They seem not to be solitons because numerical calculations indicate that they do not retain their identies after interactions with one another. (Stern (1975); Larichev & Reznik (1976); McWilliams & Zabusky (1982).)]

2.15 *Another form of the cubic Schrödinger equation.* Consider the equation

$$iu_t + u_{xx} - |u|^2u = 0$$

and seek waves of permanent form by trying the solution

$$u = re^{-i(\theta + \omega t)},$$

where r and θ are real functions of $X = x - ct$ only and where ω and c are real constants. Show that, in particular, there exist localized waves with

$$r^2(X) = \omega - 4k^2 \operatorname{sech}^2 kX$$

and

$$c\tan(2k\theta/c) = 2\sqrt{2}k\tanh kX$$

for all c such that $k = \frac{1}{2}\sqrt{\omega - \frac{1}{2}c^2} > 0$.

[Zakharov & Shabat (1973). This solution, like those of Problems 2.5 and 2.10, takes the form of a modulated carrier wave, i.e. an oscillation within an envelope. They indicate the variety of forms a soliton may have. The solution of this problem is applicable to nonlinear optics.]

CHAPTER 3 CONSERVATION LAWS

1 *Fundamental ideas*

Suppose that T and X are some functions of u, the spatial derivatives of u, x and t so that

$$\frac{\partial T}{\partial t} + \frac{\partial X}{\partial x} = 0 \tag{1}$$

whenever u(x,t) is a solution of some given equation, for example the KdV equation. Then we call the equation displayed above a *conservation law* or *conservation relation* with *density* T and *flux* X. You may be familiar with the special case of the conservation of mass of a compressible fluid of density ρ and mass flux ρu in the x-direction, governed by the continuity equation

$$\frac{\partial \rho}{\partial t} + \frac{\partial (\rho u)}{\partial x} = 0.$$

There are similar equations for the conservation of electric charge and other quantities.

If T and X are integrable for $-\infty < x < \infty$, then

$$\frac{d}{dt}\int_{-\infty}^{\infty} T\, dx = -\,[X]_{-\infty}^{\infty} = 0.$$

Therefore

$$\int_{-\infty}^{\infty} T\, dx = \text{constant}. \tag{2}$$

These ideas are applicable to many equations and can express the conservation of many physical quantities. For the KdV equation one can easily see that

$$0 = u_t - 6uu_x + u_{xxx}$$

$$= \frac{\partial u}{\partial t} + \frac{\partial}{\partial x}(-3u^2 + u_{xx}). \qquad (3)$$

Therefore one choice of density and flux is given by

$$T = u \quad \text{and} \quad X = u_{xx} - 3u^2. \qquad (4)$$

It follows that

$$\int_{-\infty}^{\infty} u dx = \text{constant}, \qquad (5)$$

where u is any solution of the KdV equation. Also

$$0 = u(u_t - 6uu_x + u_{xxx})$$

$$= \frac{\partial}{\partial t}(\tfrac{1}{2}u^2) + \frac{\partial}{\partial x}(-2u^3 + uu_{xx} - \tfrac{1}{2}u_x^2).$$

Therefore similarly

$$\int_{-\infty}^{\infty} u^2 dx = \text{constant}. \qquad (6)$$

The facts that the above two integrals of the densities are constants of the motion represent respectively the conservation of mass and momentum of the water waves described by the KdV equation. (Remember that u represents the depth of the surface of the water.)

*There is a deeper approach to this topic. Conservation laws for many systems are well-known to arise from variational principles when the Lagrangian density is invariant under transformations belonging to a continuous group, e.g. translations in space, translations in time, Galilean transformations, and rotations. A conservation law follows from an invariant Lagrangian density by Noether's theorem. For further reading about this approach, the book by Gel'fand & Fomin (1963, §§37,38) is recommended.

2 *Gardner's transformation*

Hitherto we have found conservation laws by trial and error. Next we shall use some ideas of Miura, Gardner & Kruskal (1968) to generate conservation laws systematically.

Define w by the relation

$$u = w + \varepsilon w_x + \varepsilon^2 w^2 \tag{1}$$

in terms of any well-behaved function u, where ε is any real parameter. (Of course, w is not uniquely defined in this way.) Then

$$\begin{aligned} & u_t - 6uu_x + u_{xxx} \\ &= w_t + \varepsilon w_{xt} + 2\varepsilon^2 ww_t - 6(w + \varepsilon w_x + \varepsilon^2 w^2)(w_x + \varepsilon w_{xx} + 2\varepsilon^2 ww_x) \\ &\quad + w_{xxx} + \varepsilon w_{xxxx} + 2\varepsilon^2 (ww_x)_{xx} \\ &= (1 + \varepsilon\frac{\partial}{\partial x} + 2\varepsilon^2 w)\{w_t - 6(w + \varepsilon^2 w^2)w_x + w_{xxx}\}. \end{aligned} \tag{2}$$

Therefore the KdV equation is satisfied by u if w satisfies the *Gardner equation*,

$$w_t - 6(w + \varepsilon^2 w^2)w_x + w_{xxx} = 0, \tag{3}$$

but not necessarily vice versa (Miura, Gardner & Kruskal 1968).

Note that Gardner's equation has a conservation law of the form

$$w_t + \{- 3w^2 - 2\varepsilon^2 w^3 + w_{xx}\}_x = 0. \tag{4}$$

Next suppose that we can expand

$$w = \sum_{n=0}^{\infty} \varepsilon^n w_n(u) \tag{5}$$

for small ε. Then Gardner's transformation

$$w = u - \varepsilon w_x - \varepsilon^2 w^2$$

gives, on equating coefficients of ε^o, ε^1, ε^2, ... ,

$$w_o = u, \tag{6a}$$

$$w_1 = - w_{ox} = - u_x, \tag{6b}$$

$$w_2 = - w_{1x} - w_o^2 = u_{xx} - u^2, \; \ldots \; . \tag{6c}$$

Now we can equate coefficients of ε^n in the conservation-law form of the Gardner equation for $n = 0, 1, 2, \ldots$ to find more conservation laws for the KdV equation. Thus

$\underline{\varepsilon^o|}$ $$w_{ot} + (- 3w_o^2 + w_{oxx})_x = 0,$$

i.e. $$u_t + (- 3u^2 + u_{xx})_x = 0. \tag{7}$$

$\underline{\varepsilon^1|}$ $$(w_1)_t + (- 6w_o w_1 + w_{1xx})_x = 0,$$

i.e. $$(- u_x)_t + (6uu_x - u_{xxx})_x = 0.$$

This conservation law is trivial, being essentially the derivative of the KdV equation with respect of x. It gives

$$\text{constant} = \int_{-\infty}^{\infty} u_x dx = [u]_{-\infty}^{\infty} = \text{constant}$$

if u tends to constant values as $x \to \pm\infty$.

$\underline{\varepsilon^2|}$ $$w_{2t} + (- 3w_1^2 - 6w_o w_2 - 2w_o^3 + w_{2xx})_x = 0,$$

i.e. $$(u_{xx} - u^2)_t + \{- 3u_x^2 - 6u(u_{xx} - u^2) - 2u^3 + u_{xxxx} - (2uu_x)_x\}_x$$

$$= 0,$$

i.e. $$(u_{xx} - u^2)_t + (4u^3 - 8uu_{xx} - 5u_x^2 + u_{xxxx})_x = 0. \tag{8}$$

We see that equation (7) is the same as (1.3) and that equation (8) gives (1.6). On equating coefficients of powers of ε, it can be seen that the odd powers give no useful information but that the even powers give independent conservation laws for the KdV equation. The details of this process are less important than its conclusion that there is an *infinity* of densities conserved. This and the numerical results of §1.3 suggest plausibly that if a lot of separated solitary waves travel to meet one another, interact and separate then they will conserve their identities in order that all the integrated densities remain constants of the motion.

Problems

3.1 *Conservation law for the Burgers equation.* Show that if

$$u_t + uu_x = \nu u_{xx}$$

then u is a density in a conservation law.

3.2 *Conservation laws for the KdV equation.* Show that $u^3 + \frac{1}{2}u_x^2$ and $xu + 3tu^2$ are densities in conservation laws for the KdV equation in standard form. What are the corresponding fluxes? Find another density that is new to you.

[Miura, Gardner & Kruskal (1968).]

3.3 *Conservation laws for a modified KdV equation.* Find at least two independent conservation laws for the modified KdV equation,

$$u_t - 6u^2u_x + u_{xxx} = 0.$$

[Miura (1974, p.218).]

3.4 *Conservation laws for the RLW equation.* Show that the RLW equation has the equivalent form,

$$u_t - uu_x - u_{xxt} = 0.$$

Then deduce the conservation laws,

$$u_t - (u_{xt} + \tfrac{1}{2}u^2)_x = 0,$$

$$\tfrac{1}{2}(u^2 + u_x^2)_t - (uu_{xt} + \frac{1}{3}u^3)_x = 0,$$

and

$$\left(\frac{1}{3}u^3\right)_t + (u_t^2 - u_{xt}^2 - u^2u_{xt} - \tfrac{1}{4}u^4)_x = 0.$$

[There are, in fact, no more non-trivial independent conservation laws in which the density depends smoothly upon only u and its spatial derivatives (Olver 1979).]

*3.5 *Conservation laws for the DABO equation.* Show that if

$$u_t + uu_x + \{\mathscr{H}(u)\}_{xx} = 0$$

and $u, u_x \to 0$ sufficiently rapidly as $x \to \pm\infty$, then $\int_{-\infty}^{\infty} u\,dx$, $\int_{-\infty}^{\infty} u^2 dx$, $\int_{-\infty}^{\infty}\left\{\frac{1}{3}u^3 - u_x\mathscr{H}(u)\right\}dx$, and $\frac{d}{dt}\int_{-\infty}^{\infty} xu\,dx$ are independent of time.

[See Problem 2.9; Ono (1975).]

3.6 *A variational principle for the KdV equation.* Deduce the KdV equation by considering the variational principle $\delta\int \mathscr{L}dx\,dt = 0$ with the Lagrangian density

$$\mathscr{L} = \tfrac{1}{2}v_x v_t - v_x^3 - \tfrac{1}{2}v_{xx}^2$$

and identifying $u = v_x$. Show that $\mathscr{L}$ is invariant under translation of x, t and v.

Hence find the Hamiltonian, $\mathscr{H} = v_t \partial\mathscr{L}/\partial v_t - \mathscr{L}$. Deduce that the total 'energy' $\int_{-\infty}^{\infty}\mathscr{H}dx = \int_{-\infty}^{\infty}(u^3 + \tfrac{1}{2}u_x^2)dx$.

[Gardner (1971).]

3.7 *Conservation laws for the cubic Schrödinger equation.* Show that if

$$iu_t + u_{xx} + \nu|u|^2 u = 0 \qquad \text{for} \quad -\infty < x < \infty$$

then $\int_{-\infty}^{\infty}|u|^2 dx$, $\int_{-\infty}^{\infty} i(\bar{u}\,u_x - u\,\bar{u}_x)dx$ and $\int_{-\infty}^{\infty}(|u_x|^2 - \tfrac{1}{2}\nu|u|^4)dx$ are constant, where the overbars denote complex conjugates.

[Zakharov & Shabat (1972).]

3.8 *The Born-Infeld equation.* Use the variational principle $\delta\int\mathscr{L}dxdt = 0$ with the Lagrangian density

$$\mathscr{L} = \sqrt{1 - u_t^2 + u_x^2}$$

to deduce that

$$(1 - u_t^2)u_{xx} + 2u_x u_t u_{xt} - (1 + u_x^2)u_{tt} = 0.$$

Show that $u(x,t) = f(x \pm t)$ gives two solutions for any twice differentiable function f.

[Born & Infeld (1934) proposed a three-dimensional form of the equation to model a relativistic particle. See also Barbashov & Chernikov (1967).]

3.9 *Conservation of quasi-geostrophic potential vorticity.* As in Problem 2.14, you are given that

$$\zeta_t - R^{-2}\psi_t + \beta\psi_x + \psi_x\zeta_y - \psi_y\zeta_x = 0,$$

where $\zeta = \psi_{xx} + \psi_{yy}$.

Show that

$$q_t - \psi_y q_x + \psi_x q_y = 0,$$

where the quasi-geostrophic potential vorticity q is defined as $q = \zeta - \psi/R^2 + \beta y$. Deduce that

$$F_t - \psi_y F_x + \psi_x F_y = 0$$

for all differentiable functions $F(q)$.

Hence or otherwise show that $\iint_{\mathbb{R}^2} F(q)\,dxdy$ is a constant density for all smooth solutions ψ which vanish sufficiently rapidly in the plane at infinity.

CHAPTER 4 THE INITIAL-VALUE PROBLEM FOR THE KORTEWEG-DE VRIES EQUATION

1 *The problem*

Next we shall reveal the remarkable role of solitons by solving the initial-value problem for the KdV equation over the interval $(-\infty, \infty)$, i.e. by finding the solution $u(x, t)$ of

$$u_t - 6uu_x + u_{xxx} = 0 \qquad (1)$$

for all $t > 0$ and $-\infty < x < \infty$, where

$$u(x, 0) = g(x) \qquad (2)$$

for a given function g. We may require that

$$\int_{-\infty}^{\infty} \left|\frac{d^n g}{dx^n}\right|^2 < \infty \quad \text{for } n = 0, 1, 2, 3, 4,$$

in order to prove that there exists a unique smooth solution of the initial-value problem (Bona & Smith 1975). These conditions seem not to be necessary ones, and, indeed, we shall violate them in §8 and find what is believed to be a smooth unique solution for $t > 0$. We also require that

$$\int_{-\infty}^{\infty} (1 + |x|)\,|g(x)|\,dx < \infty$$

in order that a certain eigenvalue problem in §3 has a solution (Faddeev 1958). Not emphasizing rigour, we shall not discuss these conditions further.

2 *Sketch of the method of inverse scattering*

The *method of inverse scattering* or the *inverse scattering transform* will be used to solve the initial-value problem. The method is quite involved so we shall sketch the structure of the method before drawing the details.

To solve the initial-value problem, we shall introduce an auxiliary or transformed problem, namely a linear one-dimensional scattering problem in which the time t of the KdV equation appears only as a parameter. The initial solution g of the KdV equation is the scattering 'potential' which gives the scattering 'data', i.e. the discrete eigenvalues and the reflection coefficient of the scattering problem, at $t = 0$. With the aid of an ingeniously chosen evolution equation we shall then deduce quite simply the scattering data for $t > 0$ in terms of the scattering data for $t = 0$. Finally, we shall invert the scattering data to find the 'potential' for $t > 0$ and identify the potential with the solution $u(x, t)$ of the KdV equation. The method takes us a long way round to solve the nonlinear equation and seems to be tractable because the scattering problem is not only an ordinary differential system but also a linear one. The final solution, when all is done, is not easy to untangle, but it does give qualitative as well as quantitative results.

It may help to understand the advantage of this circuitous route by considering the (incomplete) analogy of the well-known use of the Laplace transform to solve linear initial-value problems. We often solve a relatively difficult partial differential equation with independent variables x and t by first taking its Laplace transform with respect

Fig. 1. Flow diagram of the method of inverse scattering.

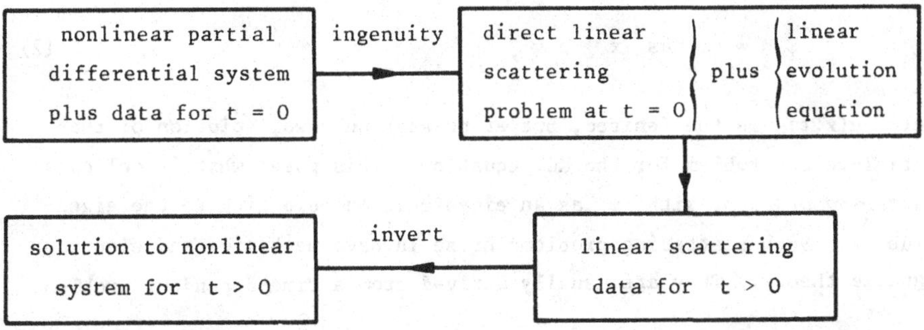

to t. This gives an easier ordinary differential equation with independent variable x. We solve this for all x using the initial data at $t = 0$. Finally we invert the transform to find the solution of the partial differential equation for all $t > 0$ and x.

The method of inverse scattering is illustrated symbolically in the flow diagram of Fig. 1.

We shall next indicate how Kruskal and his associates (Gardner, Greene, Kruskal & Miura 1967) invented this method although their motivation is now largely of historical interest. The numerical results known to them indicated what had to be proved. They were influenced by their knowledge of the solution of the *Burgers equation* (see Problem 4.2). Also *Miura's transformation* related Riccati's equation to the KdV equation (Problem 4.3). Yet the chief justification of the method was that it worked. The crucial steps are the choices of the linear ordinary differential equation of the scattering problem and of the evolution equation. These choices were, perhaps, more a matter of art than science. Later, however, Lax (1968) found a more general rationale for them and paved the way for the application of the method of inverse scattering to several nonlinear partial differential equations. Lax's method will be explained in the next chapter. A method even more useful than Lax's was devised by Ablowitz et al. (1973) and will be introduced in §6.6.

3 *The scattering problem*

Following the method sketched above, we consider $\psi(x, t)$ and $\lambda(t)$ such that

$$\psi_{xx} + (\lambda - u)\psi = 0 \tag{1}$$

and

$$\psi < \infty \quad \text{as} \quad x \to \pm\infty, \tag{2}$$

where $u(x,t)$ is the desired, but at present unknown, solution of the initial-value problem for the KdV equation. This poses what is called a *scattering problem*, with ψ as an eigenfunction belonging to the eigenvalue λ. Such scattering problems arise in wave mechanics and electromagnetic theory. They are usually derived from a time-dependent problem, for example Schrödinger's equation or Maxwell's equations, but that time is entirely independent of the variable t of the KdV equation. In our scattering problem, t merely appears as a passive parameter. The reason for considering the scattering problem will soon emerge: $\lambda(t)$ develops in a very simple way while the solution $u(x,t)$ of the KdV equation develops in a complicated way.

The well-known theory of the scattering problem may be found in many textbooks. If you are not familiar with the elements of scattering theory, it is recommended that you read Landau & Lifshitz (1965, §§21-25) or your favourite textbook, although we shall first briefly show how the scattering problem arises from a wave equation and then develop those results which we need.

Starting with the wave equation,

$$\frac{\partial^2\psi}{\partial x^2} - u\psi = \frac{1}{c^2}\frac{\partial^2\psi}{\partial\tau^2}, \tag{3}$$

we may seek solutions with frequency ω so that $\psi \propto e^{-i\omega\tau}$, and recover equation (1) if $\lambda = \omega^2/c^2$. In regions where the 'potential' field u vanishes, there may exist solutions $\psi(x,t) = \mathrm{Re}\{a\, e^{-i(kx+\omega\tau)}\}$ and $\psi(x,t) = \mathrm{Re}\{b\, e^{i(kx-\omega\tau)}\}$, where $k = \omega/c$, and a and b are complex constants. Taking ω and k to be positive without loss of generality, we associate $a\, e^{-i(kx+\omega\tau)}$ with a wave of speed c in the negative x-direction and $b\, e^{i(kx-\omega\tau)}$ with a wave of speed c in the positive x-direction. In physical applications of this wave equation the waves usually have energy fluxes proportional to $|a|^2$ and $|b|^2$ respectively. Our present work comes from the KdV equation not from the wave equation, so we may use the concepts, terminology and experience of solving the wave equation to help us, although the wave equation is not physically relevant to the problem in hand.

Next note that we assumed in §1 that $u \to 0$ sufficiently rapidly as $x \to \pm\infty$, so that $\psi'' \sim -\lambda\psi$ and ψ is asymptotically a linear combination of $\exp\{\pm i\sqrt{\lambda}x\}$. Therefore ψ decays exponentially at infinity if $\lambda < 0$ and ψ oscillates sinusoidally at infinity if $\lambda > 0$. The eigensolutions are accordingly of two kinds.

(i) *Bound states*. If $\lambda < 0$ then we may define $\kappa = \sqrt{-\lambda} > 0$. Therefore $\psi(x) \sim \alpha e^{\kappa x}$ as $x \to -\infty$ for some constant α in order that ψ is not unbounded. This solution of equation (1) is such that

$$\psi(x) \sim \beta e^{\kappa x} + \gamma e^{-\kappa x} \qquad \text{as} \quad x \to \infty,$$

where the constants β and γ are proportional to α and depend on κ and u. For only special values of λ, namely eigenvalues, will $\beta = 0$ and ψ be bounded as $x \to +\infty$. For these eigenvalues, the eigenfunction $\psi \to 0$ as $x \to \pm\infty$.

If κ^2 is an upper bound of $-u$ then $\psi_{xx}/\psi = \kappa^2 + u > 0$ and ψ is exponential in character for all x and cannot vanish at two points. Therefore $0 < \kappa^2 < -u_m$ for all eigensolutions of equation (1) and boundary conditions (2), where $u_m = \min_{-\infty<x<\infty} u$. As $\kappa^2 = -\lambda$ decreases from $-u_m$ the oscillations of ψ become more rapid so that β and ψ may have successive zeros, as is given by Sturm's theory of oscillations. It follows that there is a finite number, p say, of discrete eigenvalues λ_n belonging to square-integrable eigenfunctions ψ_n. It is convenient to order the eigenvalues so that $u_m < \lambda_1 < \lambda_2 < \dots < \lambda_p < 0$ and ψ_n has $n - 1$ finite zeros. The integer p depends upon u, of course; and, indeed, $p = 0$ certainly if $u(x,t) > 0$ for all x and so $u_m \geq 0$.

It is convenient to normalize ψ_n so that

$$\psi_n(x) \sim \exp(-\kappa_n x) \qquad \text{as} \quad x \to \infty. \tag{4}$$

This means that we choose the constant γ above to be equal to one for each eigenfunction. (An equally acceptable alternative would be to normalize so that $\int_{-\infty}^{\infty} \psi_n^2 dx = 1$.)

Fig. 2. Symbolic sketch of scattering by a potential.

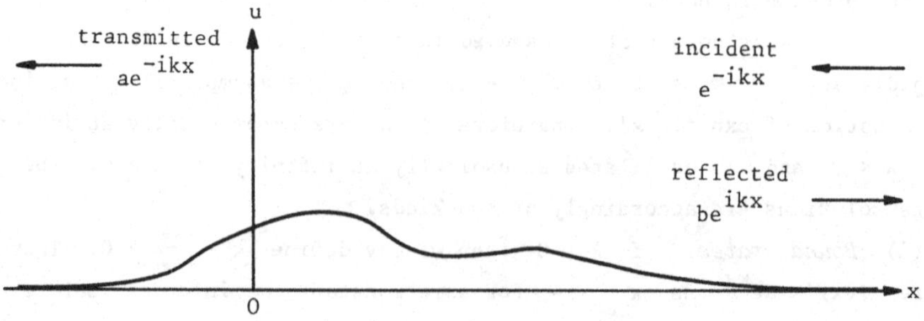

(ii) *Unbound states.* If $\lambda > 0$ then all solutions ψ of equation (1) are sinusoidal as $x \to \pm\infty$, so that conditions (2) of boundedness are automatically satisfied. Therefore there exist eigenfunctions for all $\lambda > 0$, ψ being bounded but not square integrable.

It is conventional to express any solution as a superposition of those solutions which represent an incident wave (from $x = \infty$, say) together with a reflected wave and a transmitted wave. This representation follows naturally from the wave equation (3). In mathematical terms we set $k = \sqrt{\lambda} > 0$ and seek ψ, where

$$\psi(x,t) \sim \begin{cases} e^{-ikx} + b(k,t)e^{ikx} & \text{as } x \to \infty \\ a(k,t)e^{-ikx} & \text{as } x \to -\infty \end{cases} \tag{5}$$

for some complex transmission coefficient a and reflection coefficient b. This defines the solution ψ uniquely, and we can in principle calculate a and b.

The one-dimensional scattering of an incident wave by a 'potential' u is represented symbolically in Fig. 2.

4 *The evolution equation*

Next we use the scattering equation to eliminate u partially in favour of ψ in the KdV equation, and thereby deduce the remarkable result that λ is independent of t.

Rather than substitute $u = \lambda + \psi_{xx}/\psi$ into the KdV equation directly, we find it neater to do it indirectly by defining

$$R = \psi_t + u_x\psi - 2(u + 2\lambda)\psi_x. \tag{1}$$

Therefore

$$\psi R_x - \psi_x R = \psi^2(R/\psi)_x$$

$$= \psi\psi_{xt} - \psi_x\psi_t + \psi^2 u_{xx} - 2u_x\psi\psi_x - 2(u + 2\lambda)(\psi\psi_{xx} - \psi_x^2). \tag{2}$$

Therefore

$$\frac{\partial}{\partial x}(\psi R_x - \psi_x R) = \psi\psi_{xxt} + \psi_x\psi_{xt} - \psi_{xx}\psi_t - \psi_x\psi_{xt} + \psi^2 u_{xxx}$$

$$+ 2\psi\psi_x u_{xx} - 2u_x\psi_x^2 - 2u_x\psi\psi_{xx} - 2u_{xx}\psi\psi_x$$

$$- 2u_x(\psi\psi_{xx} - \psi_x^2) - 2(u + 2\lambda)(\psi_x\psi_{xx} + \psi\psi_{xxx}$$

$$- 2\psi_x\psi_{xx})$$

$$= \psi^2\left(\frac{\psi_{xx}}{\psi}\right)_t + \psi^2 u_{xxx} - 4u_x\psi\psi_{xx}$$

$$- 2(u + 2\lambda)(\psi\psi_{xxx} - \psi_x\psi_{xx})$$

$$= \psi^2(u - \lambda)_t + \psi^2 u_{xxx} - 4u_x\psi^2(u - \lambda)$$

$$- 2(u + 2\lambda)\psi^2\left(\frac{\psi_{xx}}{\psi}\right)_x$$

$$= \psi^2(u_t - \lambda_t + u_{xxx} - 6uu_x).$$

It follows that if $\psi_{xx} + (\lambda - u)\psi = 0$ and $u_t - 6uu_x + u_{xxx} = 0$ then

$$\psi^2\lambda_t + \frac{\partial}{\partial x}(\psi R_x - \psi_x R) = 0. \qquad (3)$$

Consider the bound states here. Then $\lambda = \lambda_n < 0$ and $\psi = \psi_n$ where $\int_{-\infty}^{\infty}\psi_n^2 dx < \infty$. Therefore we can form

$$\int_{-\infty}^{\infty}\psi^2 dx\lambda_t = -[\psi R_x - \psi_x R]_{-\infty}^{\infty} = 0,$$

on dropping the subscripts n for convenience. Therefore

$$\lambda_t = 0, \tag{4}$$

i.e. the eigenvalues λ_n are independent of time. Therefore

$$\frac{\partial}{\partial x}(\psi R_x - \psi_x R) = 0. \tag{5}$$

Therefore

$$\psi R_x - \psi_x R = \text{function of time only}$$

$$= \text{value at } x = \infty$$

$$= 0,$$

i.e.

$$\psi^2 \frac{\partial}{\partial x}\left(\frac{R}{\psi}\right) = 0. \tag{6}$$

Therefore

$$R/\psi = \text{function of time only}$$

$$= \text{value at } x = \infty$$

$$= \left[\frac{\psi_t + u_x\psi - 2(u + 2\lambda)\psi_x}{\psi}\right]_{x = \infty}$$

$$= \left[\frac{0 + 0 \times e^{-\kappa x} - 2(0 + 2\lambda)(-e^{-\kappa x})}{e^{-\kappa x}}\right]_{x = \infty}$$

$$= 4\lambda\kappa$$

$$= -4\kappa^3.$$

Then we deduce the *evolution equation* for bound states ψ_n,

$$\psi_t + u_x\psi - 2(u + 2\lambda)\psi_x + 4\kappa^3\psi = 0. \tag{7}$$

5 *Solution of the scattering problem for* $t > 0$

We can now in principle put $u(x,0) = g(x)$ and solve the scattering problem of §3 at $t = 0$. This necessitates the solution of a linear ordinary differential equation and is relatively simple, but may in practice require the use of a computer to obtain the solution in numerical form. Thus we suppose formally that we know p, κ_n and $\psi_n(x,0)$ for $n = 1,2, \ldots, p$, and also $b(k,0)$ for all $\lambda = k^2 > 0$.

For the bound states, $\lambda_t = 0$ and therefore the κ_n are constants, and so known to be equal to their initial values for all $t > 0$. Also we shall need to use

$$c_n(t) = \left\{\int_{-\infty}^{\infty} \psi_n^2\,(x,t)\,dx\right\}^{-1} \tag{1}$$

in §6. We see that

$$\frac{d}{dt}\left(c_n^{-1}\right) = \frac{d}{dt}\int_{-\infty}^{\infty} \psi_n^2\,dx$$

$$= \int_{-\infty}^{\infty} 2\,\psi\psi_t\,dx,$$

on dropping the subscript n for convenience. Now we find

$$\frac{d}{dt}\left(c_n^{-1}\right) = 2\int_{-\infty}^{\infty} \{2(u + 2\lambda)\,\psi\psi_x - u_x\psi^2 - 4\kappa^3\psi^2\}\,dx,$$

on use of the evolution equation (4.7),

$$= 2\int_{-\infty}^{\infty} \{4u\,\psi\psi_x - 2u\,\psi\psi_x - u_x\,\psi^2\}dx - 8\kappa^3 c_n^{-1}$$

$$= 2\int_{-\infty}^{\infty} \{4(\psi_{xx} + \lambda\psi)\,\psi_x + 4\lambda\psi\psi_x - (u\psi^2)_x\}dx - 8\kappa^3 c_n^{-1}$$

$$= 2\,[2\psi_x^2 + 2\lambda\psi^2 + 2\lambda\psi^2 - u\psi^2]_{-\infty}^{\infty} - 8\kappa^3 c_n^{-1}$$

$$= -\,8\kappa^3 c_n^{-1}, \tag{2}$$

on replacing the subscript n. Therefore

$$c_n(t) = c_n(0)\exp(8\kappa_n^3 t) \tag{3}$$

for all $t > 0$, and is thereby simply found.

Next consider the unbound states with $\lambda > 0$. For them we may simply fix any value of λ so that $\lambda_t = 0$, because we know that all $\lambda > 0$ give unbound states for all solutions $u(x,t)$. Therefore, as before for the bound states, we find that

$$R/\psi = \text{function of time only}$$

$$= \text{value at } x = \infty$$

$$= \left[\frac{\psi_t + u_x\psi - 2(u + 2\lambda)\psi_x}{}\right]_{x=\infty}$$

$$= \left[\frac{b_t e^{ikx} + 0 - 2(0 + 2\lambda)ik(-e^{-ikx} + be^{ikx})}{e^{-ikx} + be^{ikx}}\right]_{x=\infty}$$

$$= \left[\frac{(b_t - 4ik\lambda b)e^{ikx} + 4ik\lambda e^{-ikx}}{be^{ikx} + e^{-ikx}}\right]_{x=\infty}$$

Therefore the coefficients of $e^{\pm ikx}$ must be in equal proportion, i.e.

$$R/\psi = 4ik\lambda = 4ik^3$$

and

$$b_t = 4ik\lambda b + 4ik\lambda b$$

$$= 8ik^3 b. \qquad (4)$$

Therefore

$$b(k,t) = b(k,0)\exp(8ik^3 t) \qquad \text{for all } t > 0. \qquad (5)$$

Similarly, we find that

$$4ik^3 = R/\psi = \left[\frac{\psi_t + u_x\psi - 2(u + 2\lambda)\psi_x}{\psi}\right]_{x=-\infty}$$

$$= \left[\frac{a_t e^{-ikx} + 0 - 4\lambda(-ikae^{-ikx})}{ae^{-ikx}}\right]_{x=-\infty}$$

$$= \frac{a_t}{a} + 4ik^3.$$

Therefore

$$a_t = 0, \tag{6}$$

and so

$$a(k,t) = a(k,0) \qquad \text{for all} \quad t > 0. \tag{7}$$

In summary, using our detailed knowledge of the asymptotic behaviour of ψ as $x \to \pm\infty$, we have found κ_n, c_n and b for all $t \geq 0$.

6 *The inverse scattering problem*

The next stage is to find $u(x,t)$ from the scattering data, namely κ_n, c_n and b, for $t > 0$. An analogue of this is to find the shape of the membrane on a drum from knowledge of all the sounds it makes when beaten, or to find the mass distribution of a stretched string from knowledge of all the sounds it makes when plucked. To solve the present problem, we quote a result following from the ideas of Gel'fand & Levitan (1951). The form of their solution of the inverse scattering problem due to Marchenko (1955) and Kay & Moses (1956a) is

$$u(x,t) = -2\frac{\partial}{\partial x}K(x,x,t), \tag{1}$$

where K is the unique solution of the linear ordinary integral equation,

$$K(x,y,t) + B(x+y,t) + \int_x^{\infty} K(x,z,t)B(y+z,t)dz = 0 \quad \text{for} \quad y > x, \tag{2}$$

and B is defined by

$$B(x+y,t) = \sum_{n=1}^{p} c_n(t)\exp\{-\kappa_n(x+y)\} + \frac{1}{2\pi}\int_{-\infty}^{\infty} b(k,t)e^{ik(x+y)}dk. \tag{3}$$

(To interpret equation (1), first put $y = x$ in $K(x,y,t)$ and then differentiate with respect to x for any fixed value of t.) The derivation of this solution is sketched in the Appendix. It may also be noted that t is merely a passive parameter in the integral equation, although it appears as an independent variable in the KdV equation.

7 *Qualitative character of the solution*

The solution is now complete, although we have not learnt much about it! The solution to the initial-value problem of the nonlinear partial differential equation has been formally reduced to the solution of the linear ordinary differential equation of the scattering problem at $t = 0$ and the solution of the linear ordinary integral equation. The reduced problems are much simpler but by no means trivial. We can, however, get valuable qualitative information without solving the problem explicitly. This is first explained briefly in this section. Some detailed examples will be given in later sections.

For $t > 0$ the solution u evolves so that all the conservation laws are satisfied, so that the eigenvalues λ_n and the transmission coefficients $a(k)$ are constants of the motion, and so that the reflection coefficients $b(k)$ oscillate in accord with equation (5.5). After a long time the solution separates into two effectively independent parts. One part is a procession of solitary waves, whose properties follow from the theory of §13. The other part is a dispersive wave train, whose properties follow from the theory of the linearized KdV equation described in §1.2 and Problem 1.4.

It can be shown that the discrete spectrum gives p solitary waves, each of the form

$$-\tfrac{1}{2}c\,\mathrm{sech}^2\{\tfrac{1}{2}\sqrt{c}(x - ct - x_o)\},$$

so that the full solution satisfies

$$u(x,t) \sim -2\sum_{n=1}^{p} \kappa_n^2 \mathrm{sech}^2\{\kappa_n(x - 4\kappa_n^2 t - x_n)\} \qquad \text{as} \quad t \to \infty. \qquad (1)$$

The phase lag,

$$x_n = \frac{1}{2\kappa_n} \ln\left\{\frac{c_n(0)}{2\kappa_n} \prod_{m=1}^{n-1} \left(\frac{\kappa_n - \kappa_m}{\kappa_n + \kappa_m}\right)^2\right\}, \qquad (2)$$

follows from the analysis of §13 (see Gardner et al. (1974, p.122); Zakharov (1971); Wadati & Toda (1972)). We interpret this so that

$$x_1 = \frac{1}{2\kappa_1} \ell n \left\{\frac{c_1(0)}{2\kappa_1}\right\}. \tag{3}$$

(Note that $c = 4\kappa_n^2$ is *not* $c_n(t)$.) With the present form of the KdV equation, these solitons emerge in the limit as $t \to \infty$ as a train of depressions moving to the right. Their amplitudes $-2\kappa_n^2$ are proportional to their velocities c so that the bigger ones move away faster and the troughs of the waves lie in a line. (In fact, Scott Russell observed trains of solitary waves of elevation in his laboratory experiments.)

The continuous spectrum is associated with dispersive wave components of the solution. These travel slowly to the left, they spread out, and they oscillate so that u is positive and negative in different places at any time. It can be shown that the amplitude of this dispersive component dies away in accordance with the linear theory like $O(t^{-1/3})$ as $t \to \infty$ (see Segur (1973) and Problem 1.4), leaving only the solitons in a procession far to the right.

Fig. 3. A sketch of a solution for (a) $t = 0$ and (b) t large.

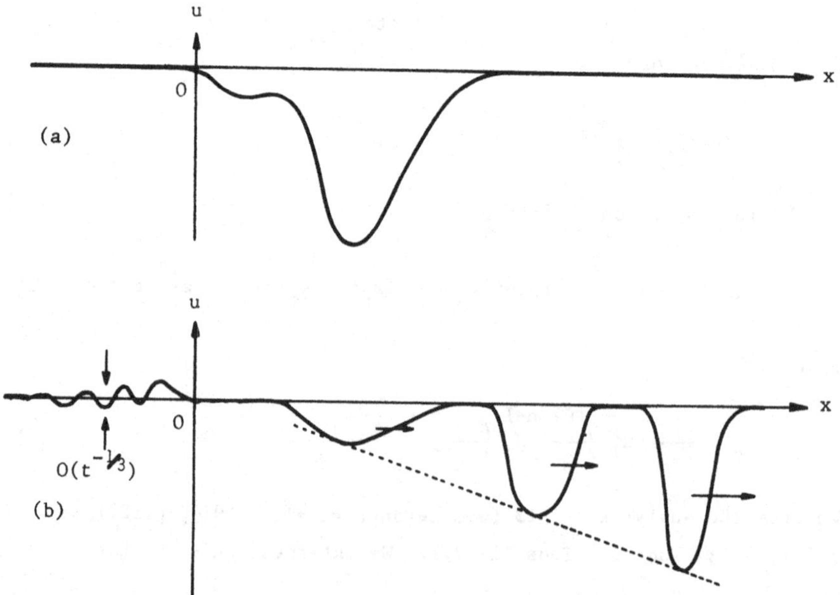

8 *Example: the delta-function potential*

Suppose that the initial form of u is

$$g(x) = -V\delta(x) \qquad \text{for } -\infty < x < \infty \tag{1}$$

and seek $u(x,t)$ for $t > 0$, where δ is Dirac's delta function.

First, we have to solve the direct scattering problem at $t = 0$:

$$\psi_{xx} + \{\lambda + V\delta(x)\}\psi = 0.$$

We anticipate that ψ_{xx} behaves like $\delta(x)$, ψ_x behaves like $\int\delta(x)dx$, i.e. like a Heaviside function, and ψ is continuous at $x = 0$. These results will be verified after the event. They give

$$[\psi_x]_{-\varepsilon}^{\varepsilon} = -\int_{-\varepsilon}^{\varepsilon}\{\lambda + V\delta(x)\}\psi\, dx$$

$$\to -V\psi(0) \qquad \text{as } \varepsilon \downarrow 0, \tag{2}$$

and the continuity of ψ at $x = 0$.

Next assume that $\lambda < 0$ in order to find the discrete spectrum. So put $\lambda = -\kappa^2$ for $\kappa > 0$. Then

$$\psi_{xx} - \kappa^2\psi = 0 \qquad \text{for } x > 0 \text{ and } x < 0. \tag{3}$$

Therefore

$$\psi(x) = \begin{cases} e^{-\kappa x} & \text{for } x > 0 \\ e^{\kappa x} & \text{for } x < 0 \end{cases}, \tag{4}$$

on solving the equation, using the continuity of ψ at $x = 0$, and normalizing so that $\psi(x) \sim 1 \times e^{-\kappa x}$ as $x \to \infty$. Finally, the other condition at $x = 0$ gives

$$-\kappa - \kappa = [\psi_x]_{0-}^{0+} = -V\psi(0) = -V.$$

Therefore

$$p = 1, \quad \kappa_1 = \tfrac{1}{2}V \quad \text{and} \quad \lambda_1 = -\tfrac{1}{4}V^2 \tag{5}$$

for all $V > 0$. Also

$$c_1(0) = \left\{\int_{-\infty}^{\infty} \psi_1^2 dx\right\}^{-1} = \left\{2\int_0^{\infty} \exp(-2\kappa_1 x)dx\right\}^{-1}$$

$$= \kappa_1 = \tfrac{1}{2}V. \tag{6}$$

The discrete spectrum is empty if $V \leqslant 0$, because we needed to take $\kappa > 0$ in order that $\psi \to 0$ as $x \to \infty$.

To find the unbound states, let $\lambda = k^2$ for $k > 0$. Therefore it follows that

$$\psi(x) = \begin{cases} e^{-ikx} + b(k,0)e^{ikx} & \text{for } x > 0 \\ \{1 + b(k,0)\}e^{-ikx} & \text{for } x < 0 \end{cases} \tag{7}$$

in order to satisfy the equation for all $x \neq 0$, the boundary conditions at infinity, and the continuity of ψ at $x = 0$. It only remains to satisfy the other condition at $x = 0$,

$$-ik + ikb - (-ik)(1 + b) = [\psi_x]_{0-}^{0+} = -V\psi(0) = -V(1 + b).$$

Therefore

$$2ikb + Vb = -V.$$

Therefore

$$b(k,0) = -\frac{V}{V + 2ik}. \tag{8}$$

Next we deduce that $\lambda_1(t) = -\tfrac{1}{4}V^2$, $c_1(t) = \tfrac{1}{2}V\exp(V^3t)$ and $b(k,t) = -V\exp(8ik^3t)/(V + 2ik)$ for all $t > 0$, on using equations (4.4), (5.3) and (5.5). Thus equation (6.3) gives B explicitly. We may now in principle solve the integral equation (6.2) to find K and afterwards deduce $u(x,t)$ for $t > 0$ from equation (6.1). However, even in this seemingly simple example, we cannot solve the integral equation (6.2) in finite terms of well-known functions. We could use a computer to solve the integral equation numerically, but must perforce be satisfied here with a qualitative description of the solution.

The delta-function trough for $V > 0$ will fill at once and soon break up into a single solitary wave moving to the right and a dispersive train of waves moving to the left. The dispersive waves spread and die out like $O(t^{-1/3})$ as $t \to \infty$. So we are left with just the single solitary wave,

$$u(x,t) \sim -2\kappa_1^2 \operatorname{sech}^2\{\kappa_1(x - 4\kappa_1^2 t - x_1)\} \qquad \text{as} \quad t \to \infty, \qquad (9)$$

where $\kappa_1 = \frac{1}{2}V$ and $x_1 = (2\kappa_1)^{-1}\ln\{c_1(0)/2\kappa_1\} = -V^{-1}\ln 2$. If $V < 0$, then the delta-function peak subsides at once and develops into a dispersive wave train with no solitary wave.

Fig. 4. The solution (a) at $t = 0$ and (b) for large t.

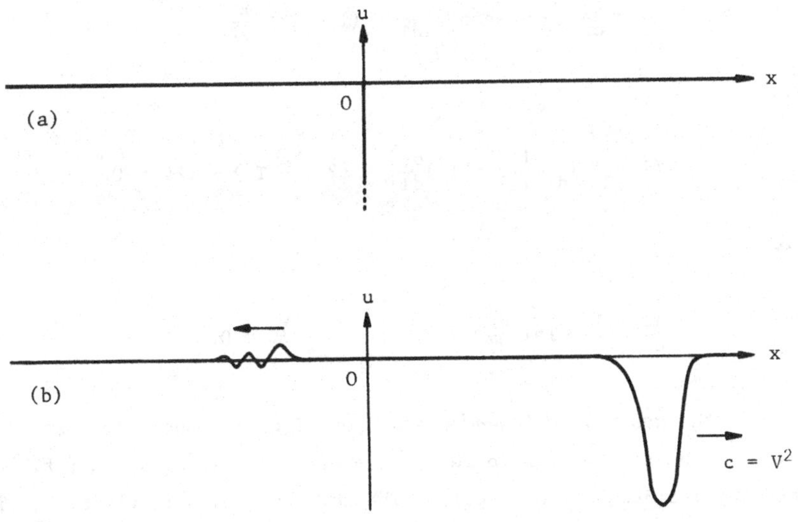

9 *Example:* $g(x) = -2\mathrm{sech}^2 x$

We shall next take the initial distribution of u to be given by

$$g(x) = -2\mathrm{sech}^2 x \qquad \text{for all } x. \tag{1}$$

We may at once recognize that the solution of the KdV equation is therefore the single solitary wave,

$$u(x,t) = -2\mathrm{sech}^2(x - 4t) \qquad \text{for all } t > 0,\ x. \tag{2}$$

The initial condition happens to represent the solitary wave as its trough passes $x = 0$. It follows that we must be able to find $p = 1$, $\lambda_1 = -1$ and $b(k,0) = 0$ for all k. We shall, however, deduce these results laboriously in order to exemplify the method of inverse scattering.

We first solve the scattering equation at $t = 0$:

$$\psi_{xx} + (2\mathrm{sech}^2 x + \lambda)\psi = 0. \tag{3}$$

It is convenient to substitute $T = \tanh x$ as the independent variable. Therefore

$$\frac{d}{dx} = \frac{dT}{dx}\frac{d}{dT} = \mathrm{sech}^2 x \frac{d}{dT} = (1 - T^2)\frac{d}{dT}$$

and

$$(1 - T^2)\frac{d}{dT}\left\{(1 - T^2)\frac{d\psi}{dT}\right\} + \{2(1 - T^2) + \lambda\}\psi = 0,$$

i.e.

$$\frac{d}{dT}\left\{(1 - T^2)\frac{d\psi}{dT}\right\} + \left(2 + \frac{\lambda}{1 - T^2}\right)\psi = 0. \tag{4}$$

This is the *associated Legendre equation* of degree one and order $\kappa = \sqrt{-\lambda}$, which may be simply transformed to a hypergeometric equation (cf. Abramowitz & Stegun 1964, §8.1). The solution which is finite at $T = 1$ (i.e. $x = \infty$) gives (Abramowitz & Stegun, §15.3.3)

$$\psi(T) \propto (1 - T^2)^{\frac{1}{2}\kappa}\, F(\kappa - 1, \kappa + 2;\ \kappa + 1;\ \tfrac{1}{2}(1 - T)), \tag{5}$$

where F is the hypergeometric function. In order that ψ is finite at $T = -1$ (i.e. $x = -\infty$) it follows (from §15.3.6) that $\kappa = 1$ and

$$\psi(T) \propto (1 - T^2)^{\frac{1}{2}}.$$

Now we normalize by convention so that

$$\psi(x) \sim e^{-\kappa x} \qquad \text{as} \quad x \to \infty$$

$$\sim \{\tfrac{1}{2}(1 - T)\}^{\frac{1}{2}\kappa} \qquad \text{as} \quad T \to 1.$$

Therefore

$$\psi(T) = \tfrac{1}{2}(1 - T^2)^{\frac{1}{2}} = \tfrac{1}{2}\operatorname{sech} x. \tag{6}$$

This gives the unique bound state. Therefore

$$\lambda_1 = -1, \tag{7}$$

and

$$c_1(0) = \left\{\int_{-\infty}^{\infty} \psi_1^2 dx\right\}^{-1} = \left\{\int_{-1}^{1} \tfrac{1}{4}(1 - T^2)\,\frac{dT}{1 - T^2}\right\}^{-1}$$

$$= 2. \tag{8}$$

Therefore

$$c_1(t) = 2e^{8t}. \tag{9}$$

For the unbound states, we put $\lambda = k^2$ for $k > 0$. Then we require that $\psi(x) \sim a\, e^{-ikx}$ as $x \to -\infty$. However, we have $1 + T = 2e^{2x}/(1 + e^{2x}) \sim 2e^{2x}$ as $x \to -\infty$, i.e. as $T \to -1$. So, replacing κ by $-ik$ and T by $-T$ in the solution (5), and normalizing, we identify

$$\psi(x) = a\{\tfrac{1}{4}(1 - T^2)\}^{-\frac{1}{2}ik} F(-ik-1, -ik+2; -ik+1; \tfrac{1}{2}(1+T))$$

$$= a\{\tfrac{1}{4}(1 - T^2)\}^{-\frac{1}{2}ik}\{\tfrac{1}{2}(1 - T)\}^{ik}$$

$$\times \frac{\Gamma(-ik+1)\Gamma(-ik)}{\Gamma(-ik-1)\Gamma(-ik+2)} F(2, -1; ik+1; \tfrac{1}{2}(1-T)),$$

by use of the connection formula (Abramowitz & Stegun 1964, §15.3.6) for the hypergeometric function, where Γ is the gamma function, because $1/\Gamma(-1) = 0$. Therefore

$$\psi(x) = a\left(\frac{1-T}{1+T}\right)^{\frac{1}{2}ik}\left(\frac{ik+1}{ik-1}\right)F(2, -1; ik+1; \tfrac{1}{2}(1-T) \tag{10}$$

because $\Gamma(z+1) = z\Gamma(z)$. Now $1 - T \sim 2e^{-x}$ as $x \to \infty$. Therefore

$$\psi(x) \sim \left(\frac{ik+1}{ik-1}\right)ae^{-ikx} \qquad \text{as} \quad x \to \infty.$$

Therefore equation (3.5) gives

$$a(k) = (ik-1)/(ik+1) \quad \text{and} \quad b(k) = 0 \quad \text{for all} \quad k. \tag{11}$$

For this special choice of $u(x,0)$ there is interference between waves reflected from the 'sides' of the 'valley' of the potential $-2\text{sech}^2 x$ which annihilate the reflected wave for all k.

It now follows that definition (6.3) gives

$$B(x+y,t) = 2e^{8t-x-y}. \tag{12}$$

Therefore equation (6.2) becomes

$$K(x,y,t) + 2e^{8t-x-y} + \int_x^\infty K(x,z,t) \times 2e^{8t-y-z}dz = 0.$$

It can now be seen that $K(x,y,t) \propto e^{-y}$, i.e.

$$K(x,y,t) = w(x,t)e^{-y}, \tag{13}$$

say, where

$$w(x,t)e^{-y} + 2e^{8t-x-y} + 2e^{8t-y}w(x,t)\int_x^\infty e^{-2z}dz = 0.$$

Therefore

$$w(x,t) = \frac{-2e^{8t-x}}{1 + 2e^{8t}[-\frac{1}{2}e^{-2z}]_x^\infty}$$

$$= \frac{-2e^{4t}}{e^{x-4t} + e^{-x+4t}}$$

$$= -e^{4t}\operatorname{sech}(x - 4t). \tag{14}$$

Therefore

$$K(x,x,t) = -e^{-x+4t}\operatorname{sech}(x - 4t). \tag{15}$$

This gives finally

$$u(x,t) = -2\frac{\partial}{\partial x}\{-e^{-x+4t}\operatorname{sech}(x - 4t)\}$$

$$= -2\operatorname{sech}^2(x - 4t). \tag{16}$$

It comes as no surprise that we recover a single solitary wave, because our initial distribution $u(x,0)$ was chosen to represent it. Note that the velocity is $c = 4\kappa_1^2 = 4$ and the phase lag $x_1 = (2\kappa_1)^{-1}\ln\{c_1(0)/2\kappa_1\} = 0$.

10 *Example:* $g(x) = -6\operatorname{sech}^2 x$

Next suppose that the initial distribution of u is given by

$$g(x) = -6\operatorname{sech}^2 x \qquad \text{for all } x. \tag{1}$$

We can solve the scattering equation by a similar method to the one used in the previous section. We find the associated Legendre equation of degree two and order κ. The solution which is finite at $T = 1$ gives

$$\psi(T) \propto (1 - T^2)^{\frac{1}{2}\kappa}F(\kappa - 2, \kappa + 3; \kappa + 1; \tfrac{1}{2}(1 - T)).$$

It follows that $p = 2$ and $\psi \propto P_2^2(T)$ or $P_2^1(T)$, and thence that

$$\kappa_1 = 2, \quad \psi_1 = \tfrac{1}{4}\mathrm{sech}^2 x \quad \text{and} \quad c_1(0) = 12 \tag{2}$$

and

$$\kappa_2 = 1, \; \psi_2 = \tfrac{1}{2}\tanh x \; \mathrm{sech}\, x \quad \text{and} \quad c_2(0) = 6\,. \tag{3}$$

The method of the previous section again shows that

$$b(k,0) = 0 \qquad \text{for all } k. \tag{4}$$

We can now find B and solve the integral equation (6.2) with a solution of the form

$$K(x,y,t) = w_1(x,t)\exp(-\kappa_1 y) + w_2(x,t)\exp(-\kappa_2 y)\,.$$

This at length gives the solution

$$u(x,t) = -12\,\frac{3 + 4\cosh(2x - 8t) + \cosh(4x - 64t)}{\{3\cosh(x - 28t) + \cosh(3x - 36t)\}^2}, \tag{5}$$

a remarkably simple solution to find for a nontrivial nonlinear partial differential problem. One can verify that $u(x,0) = -6\mathrm{sech}^2 x$.

To interpret the solution, define $X_1 = x - 4\kappa_1^2 t = x - 16t$ and $X_2 = x - 4\kappa_2^2 t = x - 4t$. Then

$$u(x,t) = -12\,\frac{3 + 4\cosh(2X_1 + 24t) + \cosh 4X_1}{\{3\cosh(X_1 - 12t) + \cosh(3X_1 + 12t)\}^2}$$

$$\sim -12\,\frac{2\exp(2X_1 + 24t)}{\{\tfrac{3}{2}\exp(-X_1 + 12t) + \tfrac{1}{2}\exp(3X_1 + 12t)\}^2}$$

as $t \to \infty$ for fixed X_1,

$$= -\frac{24}{3\left\{\tfrac{1}{2}\left[\frac{1}{\sqrt{3}}\exp(2X_1) + \sqrt{3}\exp(-2X_1)\right]\right\}^2}$$

$$= -8\operatorname{sech}^2(2X_1 - x_o), \tag{6}$$

where $x_o = \ln\sqrt{3}$. Similarly

$$u(x,t) = -12\,\frac{3 + 4\cosh 2X_2 + \cosh(4X_2 - 48t)}{\{3\cosh(X_2 - 24t) + \cosh(3X_2 - 24t)\}^2}$$

$$\sim -12\,\frac{\tfrac{1}{2}\exp(-4X_2 + 48t)}{\{\tfrac{3}{2}\exp(-X_2 + 24t) + \tfrac{1}{2}\exp(24t - 3X_2)\}^2}$$

as $t \to \infty$ for fixed X_2,

$$= -2\operatorname{sech}^2(X_2 + x_o). \tag{7}$$

Fig. 5. The interaction of two solitary waves:
(a) $t = -0.5$, (b) $t = -0.1$, (c) $u(x,0) = -6\operatorname{sech}^2 x$,
(d) $t = 0.1$, (e) $t = 0.5$. Note that $\int_{-\infty}^{\infty} u\,dx$ is constant.

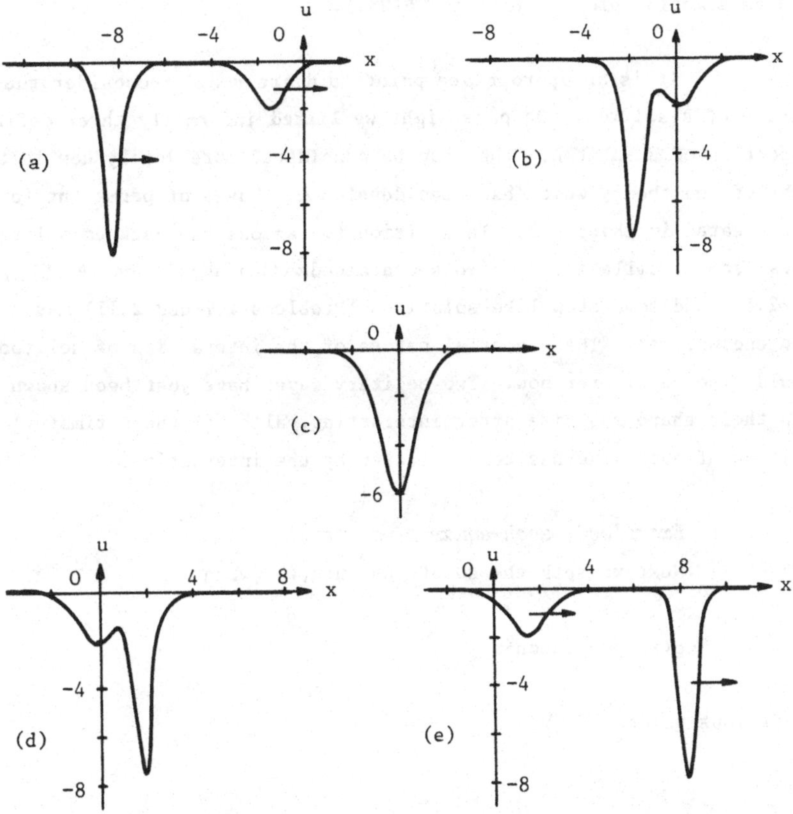

In fact, we can show similarly that

$$u(x,t) \sim - 2\text{sech}^2(X_2 \pm x_0) - 8\text{sech}^2(2X_1 \mp x_0)$$

uniformly as $t \to \pm\infty$ respectively for fixed X_1 or X_2. The remainder is exponentially small, so there is no dispersive wave train.

This solution can now be seen to represent the interaction of two solitary waves. As $t \to -\infty$ the deeper wave lies to the left of the shallower. The deeper wave catches up the shallower. They interact, and are in the middle of the interaction at $t = 0$. The deeper wave changes place with the shallower, takes the lead, and draws away. They are completely separated again as $t \to \infty$. The only ultimate vestige of the interaction is the phase shift of each solitary wave; the slower and shallower wave is retarded and the faster and deeper wave is advanced after their exchange of positions in the interaction. Some numerical results are illustrated in Fig.5. The interaction is, however, better seen as a motion picture (Eilbeck F1981).

It is an appropriate point to digress and reconsider the definition of a soliton. On page eight we listed informally three defining properties of a soliton. They can be considered more deeply now in the light of the theory which has been developed. Waves of permanent form were treated in Chapter 2. In addition to various hump-shaped solitary waves, some oscillating solutions contained within envelopes (Problems 2.5 and 2.10) and some step-like solutions (Problems 2.7 and 2.11) may have been encountered. The essential nature of the interaction of solitons should also be clearer now. Two solitary waves have just been shown to keep their shape and size after interacting, although the ultimate position of each wave has been affected by the interaction.

11 *Examples: sech-squared potentials*

Next we seek the solutions $u(x,t)$ when

$$g(x) = - V\text{sech}^2 x \tag{1}$$

for various values of V.

If $V < 0$, we already may deduce that $p = 0$, because $u(x,0) > 0$ for all x. Therefore there is no solitary-wave component of the solution as $t \to \infty$, merely a decaying dispersive wave train.

If $V = 0$, then of course the solution is $u(x,t) = 0$ for all t and x.

If $V = 2$, we may at once recognize that the solution is the single solitary wave,

$$u(x,t) = -2\operatorname{sech}^2(x - 4t) \qquad \text{for all } t > 0, x, \tag{2}$$

as shown in §9.

If $V = 6$ then there arise just two solitary waves and no dispersive wave train. They happen to be in the middle of their interaction at $t = 0$, as shown in §10.

In fact (Problem 4.5(c)), the direct scattering problem at $t = 0$ implies that if $V = \frac{1}{4}\{(2p + 1)^2 - 1\} = p(p + 1)$ for $p = 1,2 \ldots,$ then $\psi(T)$ satisfies an associate Legendre equation of degree p and order κ. Then there are p bound states with $\lambda_1 = -p^2$, $\lambda_2 = -(p-1)^2$, $\ldots$, $\lambda_p = -1$ and $b(k,0) = 0$ for all k. Also if

$$(p - 1)p < V < p(p + 1)$$

then there are p bound states with

$$\lambda_n = -(s - n)^2 \qquad \text{for } n = 1, 2, \ldots, p, \tag{3}$$

where $s = \frac{1}{2}\{\sqrt{1 + 4V} + 1\}$, and $b(k,0) \neq 0$ for some k.

It follows that if $0 < V < 2$ then there arises one solitary wave with speed $c = 4(s - 1)^2$ together with a dispersive wave train in the limit as $t \to \infty$.

If $2 < V < 6$ then there arise two solitary waves with speeds $c = 4(s - 1)^2$ and $4(s - 2)^2$ together with a dispersive wave train.

If $6 < V < 12$ then there arise three solitary waves and a dispersive wave train; and so forth.

12 *Examples: some numerical results*

In §11 we considered the solutions to the initial-value

Fig. 6. Graphs of (a) $u(x,0) = \text{sech}^2 x$, (b) $u(x,0.1)$ and (c) $u(x,0.5)$.

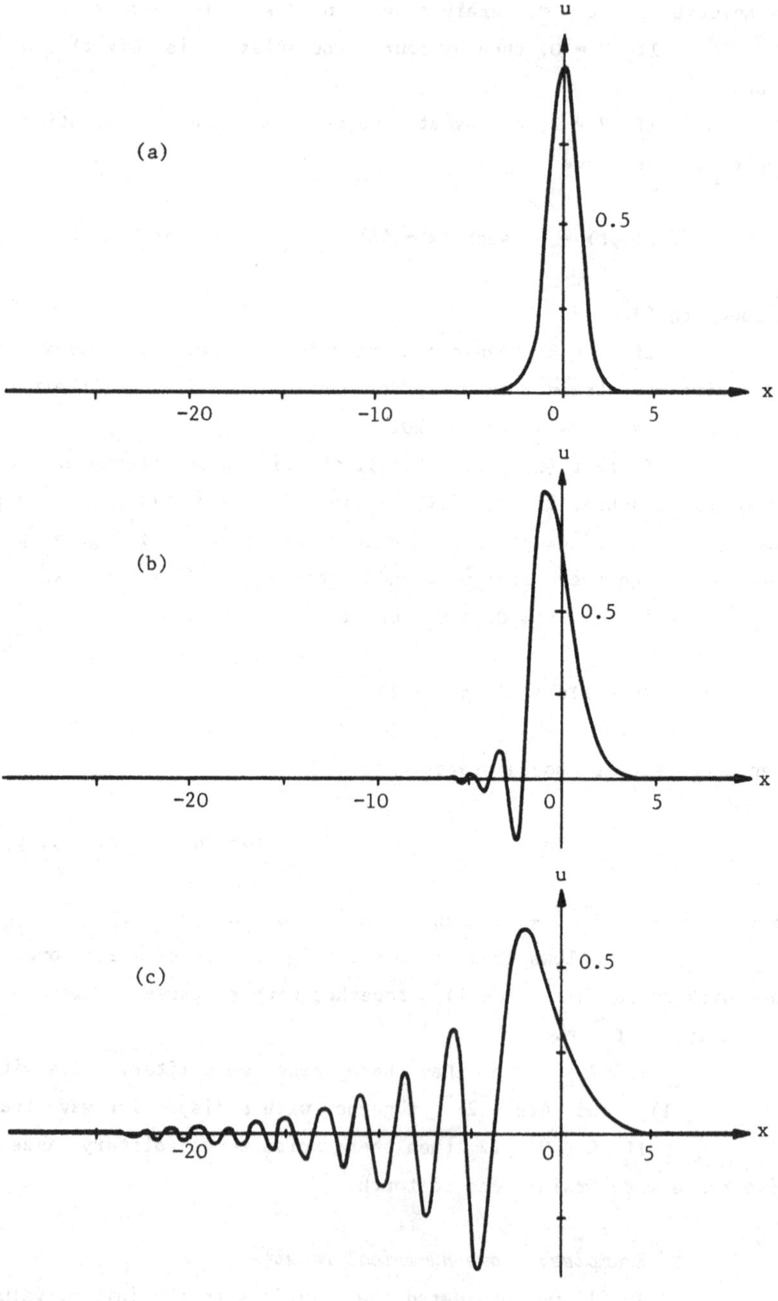

Fig. 7. Graphs of (a) $u(x,0) = -4\ \mathrm{sech}^2 x$, (b) $u(x,0.4)$ and (c) $u(x,1.0)$.

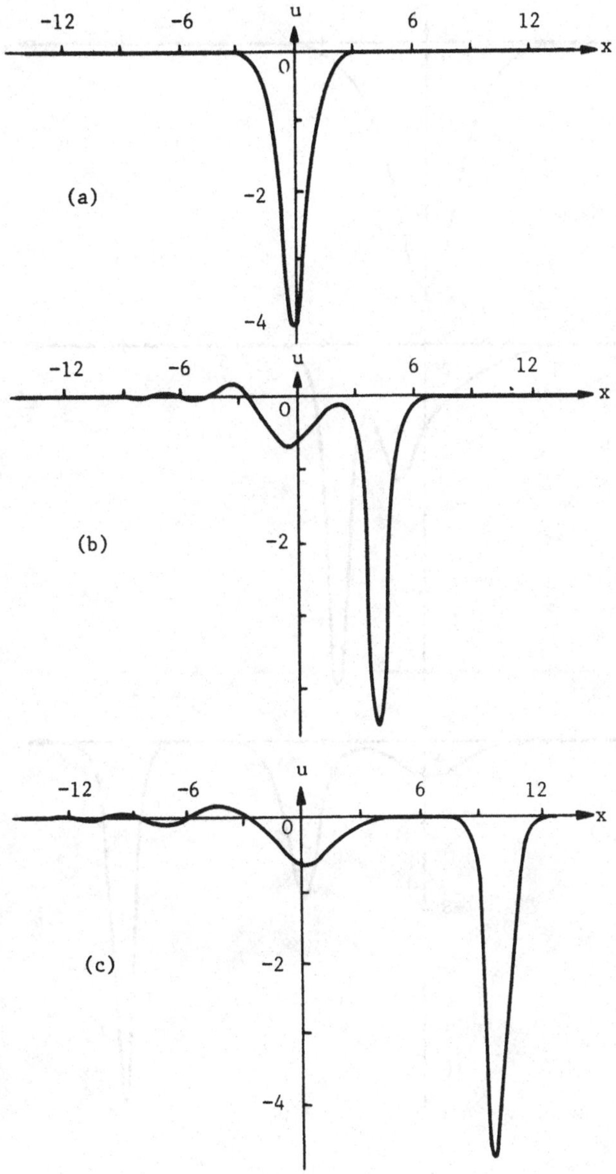

Fig. 8. Graphs of (a) $u(x,0) = -12\text{sech}^2 x$, (b) $u(x,0.05)$ and (c) $u(x,0.2)$.

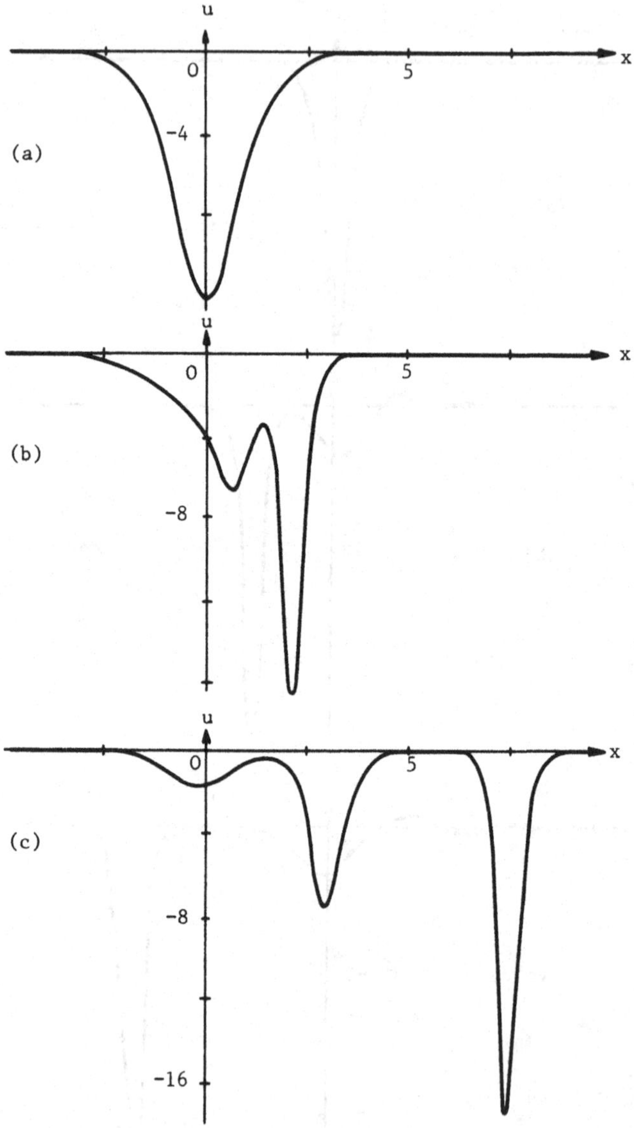

problem with

$$u(x,0) = - V\mathrm{sech}^2 x \qquad (1)$$

for all values of V. We had already found simple explicit solutions for the case of $V = 2$ in §9 and of $V = 6$ in §10. It is in general impossible to find a solution explicitly, and numerical methods must be used sooner or later. Greig & Morris (1976) have reviewed various numerical methods of directly integrating the KdV equation. Here we merely present a few results.

It can be seen that when $V = -1$ a dispersive wave train but no soliton develops as time increases. This evolution of the wave train is illustrated in Fig. 6. Note how the profile of the wave train finally resembles the graph of an Airy function. Note also that the absence of an emerging soliton is as was predicted in §11 for all $V < 0$.

It can be seen in Fig. 7 that when $V = 4$ a dispersive wave train and two solitons develop as time increases. This also is as was predicted in §11 for $2 < V < 6$.

When $V = 12$ one may use the method of §§9 and 10 to find an explicit analytic solution which represents an interaction of three solitons. They are 'caught' symmetrically at $t = 0$ and go their separate ways as $t \to \infty$ without the formation of a dispersive wave train. Some results are illustrated in Fig. 8.

We have treated the initial-value problems with $u(x,0) = - V\mathrm{sech}^2 x$ at length in §§9-12, partly because some are analytically tractable. For this reason also they provide good examples of the method of inverse scattering. But they do more. They illustrate the general qualitative character of the solutions of the initial-value problems which was discussed briefly in §7.

*13 *Reflectionless potentials*

No explicit solutions of the integral equation (6.2) have been found except when $b(k,0) = 0$ for all k. It may be surprising that b can ever be identically zero, but we have seen that $b = 0$ for some special potential functions u. (Indeed, this result is associated with the design of coating of lenses to reduce reflections of light.) In this case we now find the solution K and thence u in finite terms of elementary functions, following the analysis of Kay & Moses (1956b). If we assume that $b = 0$ then we deduce that

$$B(x + y, t) = \sum_{n=1}^{p} g_n(x,t)h_n(y), \tag{1}$$

where $g_n(x,t) = c_n(t)\exp(-\kappa_n x)$ and $h_n(y) = \exp(-\kappa_n y)$. Looking at the integral equation for $p = 1$ as in §9 and then $p = 2$ as in §10, one may see by induction that the solution is of the form

$$K(x,y,t) = \sum_{m=1}^{p} w_m(x,t)h_m(y). \tag{2}$$

It follows that the integral equation (6.2) becomes

$$\sum_{m=1}^{p} w_m(x,t)h_m(y) + \sum_{m=1}^{p} g_m(x,t)h_m(y)$$

$$+ \int_x^{\infty} \sum_{n=1}^{p} w_n(x,t)h_n(z) \sum_{m=1}^{p} g_m(z,t)h_m(y)dz = 0$$

for all $t > 0$, and for all x and y. Equating coefficients of $h_m(y) = \exp(-\kappa_m y)$, we deduce that

$$w_m + g_m + c_m \sum_{n=1}^{p} w_n \int_x^{\infty} \exp\{-(\kappa_m + \kappa_n)z\}dz = 0$$

for $m = 1, 2, \ldots, p$ and for all x, i.e.

$$w_m + g_m + c_m \sum_{n=1}^{p} w_n \frac{\exp\{-(\kappa_m + \kappa_n)x\}}{\kappa_m + \kappa_n} = 0 \tag{3}$$

i.e. $\quad Pw + g = 0,$

where w and g are $p \times 1$ column vectors with elements w_n and g_n respectively, and P is the nonsingular $p \times p$ matrix with elements

$$P_{mn}(x,t) = \delta_{mn} + c_m(t) \frac{\exp\{-(\kappa_m + \kappa_n)x\}}{\kappa_m + \kappa_n} \tag{4}$$

Therefore

$$w = -P^{-1}g.$$

Now equation (2) gives

$$K(x,x,t) = \sum_{m=1}^{p} w_m(x,t)h_m(x)$$

$$= -h^T P^{-1} g,$$

where h is the $p \times 1$ column vector with elements h_m. Also equation (4) gives

$$\frac{\partial P_{mn}(x,t)}{\partial x} = -g_m(x,t)h_n(x).$$

Therefore

$$K(x,x,t) = \text{trace}\left(P^{-1}\frac{\partial P}{\partial x}\right)$$

$$= \frac{1}{|P|}\frac{\partial |P|}{\partial x}$$

$$= \frac{\partial \ln|P|}{\partial x}, \tag{5}$$

where $|P|$ denotes det P. Finally equation (6.1) gives

$$u(x,t) = -2\frac{\partial K(x,x,t)}{\partial x}$$

$$= -2\frac{\partial^2 \ln|P|}{\partial x^2}, \tag{6}$$

the explicit solution to the KdV equation.

It can be seen at once that P is the sum of the identity matrix and a matrix whose elements are exponential functions of x and t, and so that u is expressed in finite terms of elementary functions. The fact that the solution may be evaluated explicitly in this way makes it important to recognize for a given initial distribution $u(x,0) = g(x)$ whether $b(k) = 0$ for all k. Such cases of reflectionless potentials u are very exceptional, but may be encountered (see, for example, §§9-11).

The solution (6) can be shown to give $u(x,t) < 0$ for all x and t (Problem 4.13).

Further, solution (6) can be shown at length to give

$$u(x,t) \sim -2\kappa_n^2 \operatorname{sech}^2\{\kappa_n(x - 4\kappa_n^2 t - x_n)\} \qquad \text{as} \quad t \to \infty$$

for fixed $x - 4\kappa_n^2 t$ (Gardner et al.(1974), Zakharov (1971) and Wadati & Toda (1972)), where the phase is specified by

$$\exp(2\kappa_n x_n) = \frac{c_n(0)}{2\kappa_n} \prod_{m=1}^{n-1} \left(\frac{\kappa_n - \kappa_m}{\kappa_n + \kappa_m}\right)^2 \qquad \text{for} \quad n = 1,2, \ldots, p. \tag{7}$$

It follows at once that

$$u(x,t) \sim -2 \sum_{n=1}^{p} \kappa_n^2 \operatorname{sech}^2\{\kappa_n(x - 4\kappa_n^2 t - x_n)\} \quad \text{as} \quad t \to \infty \tag{8}$$

uniformly in x. Similar expressions as $t \to -\infty$ have been found. Thus the solution represents a procession of separate solitons as $t \to \infty$, ordered according to their speeds. It is said to be a multisoliton solution because it may be interpreted as a procession of separate solitons as $t \to -\infty$, a faster one behind a slower, which pass through one another as they interact at finite values of t and finally separate in the reverse order as $t \to \infty$. The initial distribution g gives the profile of the interacting solitons at $t = 0$.

Problems

4.1 *Motion pictures of soliton interactions.* See again the animated films of solitons by Zabusky, Kruskal & Deem (F1965) and Eilbeck (F1981), and relate what you see to the theoretical results you have learnt.

4.2 *The Burgers equation.* Find a (relatively) simple well-known equation equivalent to

$$u_t + uu_x = \nu u_{xx}.$$

Hence or otherwise show that this equation does not have soliton solutions.

[This nonlinear diffusion equation was formulated by Burgers (1948) in an attempt to model turbulent flow in a channel. Hint: try $u = -2\nu v_x/v$. Forsyth (1906, p.101); Hopf (1950); Cole (1951).]

4.3 *Miura's transformation.* Show that if $u_t - 6uu_x + u_{xxx} = 0$ and u is transformed to a solution v of the Riccati equation $u = v^2 + v_x$, then

$$\left(2v + \frac{\partial}{\partial x}\right)(v_t - 6v^2 v_x + v_{xxx}) = 0.$$

Defining a function ψ so that $v = \psi_x/\psi$, deduce that

$$\psi_{xx} - u\psi = 0.$$

[It was a small step to make a Galilean transformation and consider the time-independent Schrödinger equation (4.3.1) as part of an eigenvalue problem. Miura (1968); Gardner et al. (1967).]

4.4 *Conservation of energy in direct scattering.* Show that the Wronskian $W(\psi_1,\psi_2) = \psi_1\psi_2' - \psi_1'\psi_2$ is a constant if ψ_1 and ψ_2 are each a solution of

$$\psi'' + (k^2 - u)\psi = 0$$

for given real constant k and function u such that $\int_{-\infty}^{\infty} |u| < \infty$.

Defining a solution ψ_1 by its asymptotic property that

$$\psi_1(x) \sim e^{-ikx} + b\, e^{ikx} \quad \text{as} \quad x \to \infty$$

for a given constant b, and $\psi_2 = \bar{\psi}_1$ as the complex conjugate of ψ_1, show that

$$W(\psi_1, \bar{\psi}_1) = 2ik(1 - |b|^2).$$

If, moreover,

$$\psi_1(x) \sim a e^{-ikx} \quad \text{as} \quad x \to -\infty,$$

deduce that

$$|a|^2 + |b|^2 = 1.$$

4.5 *Direct scattering problems.* Find all the bound and unbound states of the Schrödinger scattering problem, i.e. find all the discrete eigenvalues and their eigenfunctions and all the continuum of eigenvalues with the eigenfunctions, transmission and reflection coefficients for the equation,

$$\psi'' + (\lambda - u)\psi = 0 \quad \text{for} \quad -\infty < x < \infty,$$

for the following cases of potential fields.

(a) The *Dirac delta function*,

$$u(x) = -V\delta(x).$$

(b) The *rectangular well* (or *barrier*),

$$u(x) = \begin{cases} 0 & \text{for} \quad x < -1 \\ -V & \text{for} \quad -1 < x < 1 \\ 0 & \text{for} \quad 1 < x. \end{cases}$$

*(c) $\qquad u(x) = -V\operatorname{sech}^2 x;$

show here in particular that if $V > 0$ then

$$\lambda_n = -(s - n)^2 \quad \text{for} \quad n = 1, 2, \ldots, p,$$

where $s = \frac{1}{2}\{\sqrt{1 + 4V} + 1\}$ and p is the greatest integer less than s.
[Landau & Lifshitz (1965, pp. 65-7, 72-3, 78-80).]

*4.6 *Small-perturbation theory: bound states.* Taking $u(x,0) = -\epsilon f(x)$, where $0 < \int_{-\infty}^{\infty} f < \infty$, consider the eigenvalue problem that

$$\psi'' + (-\kappa^2 + \epsilon f)\psi = 0,$$

$$\psi(x) \sim e^{-\kappa x} \quad \text{as} \quad x \to \infty$$

and

$$\psi(x) \sim \gamma e^{\kappa x} \quad \text{as} \quad x \to -\infty,$$

where $\kappa > 0$ and γ is some real constant. Show that there is a unique bound state with

$$\kappa = \kappa_1 \sim \tfrac{1}{2}\epsilon \int_{-\infty}^{\infty} f$$

and

$$\left(\int_{-\infty}^{\infty} \psi^2\right)^{-1} = c_1(0) \sim \tfrac{1}{2}\epsilon \int_{-\infty}^{\infty} f \quad \text{as} \quad \epsilon \downarrow 0;$$

and that there is no bound state as $\epsilon \uparrow 0$. Show that if, however, $\int_{-\infty}^{\infty} f = 0$ then

$$\kappa_1 = \tfrac{1}{2}\epsilon^2 \int_{-\infty}^{\infty} \left(\int_x^{\infty} f(x')dx'\right)^2 dx + O(\epsilon^3) \quad \text{as} \quad \epsilon \to 0.$$

[Landau & Lifshitz (1965, p.156); Drazin (1963); Miles (1978).]

*4.7 *Small-perturbation theory: unbound states.* If $u(x,0) = -\epsilon f(x)$, then the unbound states are governed by

$$\psi'' + (k^2 + \epsilon f)\psi = 0,$$

$$\psi(x) \sim \begin{cases} e^{-ikx} + be^{ikx} & \text{as } x \to \infty \\ ae^{-ikx} & \text{as } x \to -\infty. \end{cases}$$

Deduce that

$$b(k) \sim \frac{-\varepsilon}{2ik} \int_{-\infty}^{\infty} e^{-2ikx} f(x)dx \qquad \text{as } \varepsilon \to 0.$$

[Morse & Feshbach (1953, p.1071); Drazin (1963).]

4.8 *Derivation of the densities of conservation laws from the scattering problem.* Consider the standard scattering problem for which

$$\psi_{xx} + (k^2 - u)\psi = 0$$

and

$$\psi(x) \sim e^{-ikx} + be^{ikx} \qquad \text{as } x \to +\infty.$$

for given function $u(x)$. In order to solve this problem by the JWKB method, first define h in terms of the unique solution ψ such that

$$\psi(x,k) = a(k)e^{-ikx+h(x,k)}$$

for all real x and k, where a is the transmission coefficient. Deduce that

$$h_{xx} - 2ikh_x + h_x^2 - u = 0.$$

Assuming the asymptotic expansion

$$h_x(x,k) = \sum_{n=0}^{\infty} \frac{g_n(x)}{(2ik)^{n+1}} \qquad \text{as } k \to \infty,$$

show that

$$g_0 = -u, \; g_1 = -u', \; g_2 = -u'' + u^2,$$

and

$$g_n = g'_{n-1} + \sum_{m=0}^{n-2} g_m g_{n-m-2} \quad \text{for} \quad n = 2,3,\ldots .$$

Also show that if $\mathrm{Im}(k) > 0$ then

$$\ell n(1/a) = h(\infty,k) = \sum_{n=0}^{\infty} \frac{1}{(2ik)^{n+1}} \int_{-\infty}^{\infty} g_n .$$

Deduce that $\int_{-\infty}^{\infty} g_n$ is a constant of the motion when $u(x,t)$ is an evolving solution of the standard KdV equation.

[Miura, Gardner & Kruskal (1968, p.1208).]

4.9 *An inverse scattering problem.* Show that if the discrete spectrum is empty and the reflection coefficient $b(k) = -\alpha i/(k + \alpha i)$ for $\alpha > 0$ then the solution B of equation (4.6.3) is given by

$$B(z) = -\alpha e^{\alpha z} H(-z),$$

where the Heaviside function H is defined by $H(x) = 0$ if $x \leqslant 0$ and $H(x) = 1$ if $x > 0$. Deduce that the solution $K(x,y)$ of the integral equation (4.6.2) is a step function of $-(x + y)$ and hence find the potential u.

4.10 *Some initial-value problems.* Use the method of inverse scattering to solve the initial-value problem for the KdV equation with the following initial conditions, i.e. find $u(x,t)$ for $t > 0$ and $-\infty < x < \infty$ given that

$$u_t - 6uu_x + u_{xxx} = 0 \quad \text{for} \quad t > 0 \quad \text{and} \quad -\infty < x < \infty,$$

$$\int_{-\infty}^{\infty} |u| dx < \infty \quad \text{for} \quad t > 0 \quad \text{and} \quad u(x,0) = g(x) \quad \text{for} \quad -\infty < x < \infty.$$

(a) $$g(x) = -\frac{9}{2} \operatorname{sech}^2\left(\frac{3}{2} x\right).$$

*(b) $$g(x) = \begin{cases} 0 & \text{for} \quad x < -1 \\ -V & \text{for} \quad -1 < x < 1 \\ 0 & \text{for} \quad 1 < x; \end{cases}$$

this is too difficult to solve explicitly, so just give a qualitative description of the solution $u(x,t)$ for typical values of V.

*(c) $g(x) = -V\delta(x)$; just give a qualitative description of the solution.

4.11 *Conservation of densities when two solitons interact.* Verify that

$$\int_{-\infty}^{\infty} u_1 = \int_{-\infty}^{\infty} u_2 + \int_{-\infty}^{\infty} u_3,$$

$$\int_{-\infty}^{\infty} u_1^2 = \int_{-\infty}^{\infty} u_2^2 + \int_{-\infty}^{\infty} u_3^2$$

and

$$\int_{-\infty}^{\infty} (u_1^3 + \tfrac{1}{2}u_{1x}^2) = \int_{-\infty}^{\infty} (u_2^3 + \tfrac{1}{2}u_{2x}^2) + \int_{-\infty}^{\infty} (u_3^3 + \tfrac{1}{2}u_{3x}^2),$$

where $u_1(x) = -6\mathrm{sech}^2 x$, $u_2(x) = -2\mathrm{sech}^2 x$ and $u_3(x) = -8\mathrm{sech}^2 2x$. Discuss the relationship of these results to the solution of §10.

4.12 *The uniform velocity of the 'centre of mass' of a solution.* Show that if u satisfies the KdV equation then

$$\frac{d}{dt}\int_{-\infty}^{\infty} xu\, dx = \text{constant}.$$

Interpret this result as the conservation of linear momentum of a linear mass distribution with density u. Show that this is compatible with the phase shifts of the solitary waves described in §10.

[Hint: look at Problem 3.2. Miura (1976, p.433).]

*4.13 *Reflectionless potentials.* Assume throughout this question, as in §13, that u is a reflectionless potential, i.e. that its reflection coefficient $b(k) = 0$ for all k.

Apply the operator $L_m = \partial^2/\partial x^2 - (u + \kappa_m^2)$ to equation (13.3), use equations (6.1) and (13.2), and deduce that $L_n w_n = 0$. Hence or otherwise show that

$$w_n(x,t) = -c_n(t)\psi_n(x,t),$$

where ψ_n is the usual eigenfunction belonging to eigenvalue $\lambda_n = -\kappa_n^2$

with the usual normalization so that $\psi_n(x,t) \sim \exp(-\kappa_n x)$ as $x \to \infty$.

Show also that

$$u(x,t) = 2\sum_{n=1}^{p} c_n(t)\{\psi_{nx}(x,t) - \kappa_n\psi_n(x,t)\}\exp(-\kappa_n x).$$

Eliminating $\psi_{nx} = \partial\psi_n/\partial x$ from equation (13.3) and its derivative with respect to x, deduce that

$$u(x,t) = -4\sum_{n=1}^{p} c_n(t)\kappa_n\psi_n^2(x,t).$$

Hence show that if $u(x,0)$ is a reflectionless potential and u satisfies the KdV equation then $u(x,t) < 0$ for all $t > 0$ and for all x.

[Gardner *et al.* (1974, pp.112-3).]

*4.14 *Hirota's method for multisoliton solutions.* Show that if $u_t - 6uu_x + u_{xxx} = 0$, $u = w_x$ and u and its derivatives vanish at infinity, then

$$w_t - 3w_x^2 + w_{xxx} = 0.$$

Hence or otherwise show that if, moreover, $u = -2(\ln f)_{xx}$ then

$$f(f_t + f_{xxx})_x - f_x(f_t + f_{xxx}) + 3(f_{xx}^2 - f_x f_{xxx}) = 0.$$

Substitute $f = 1 + \sum_{m=1}^{\infty}\varepsilon^m f^{(m)}$, assuming that the expansion and its derivatives converge uniformly for sufficiently small values of the parameter ε, and show that

$$\left(f_t^{(1)} + f_{xxx}^{(1)}\right)_x = 0.$$

Hence show that you may take $f^{(1)} = \sum_{j=1}^{n} f_j$ for any positive integer n, where $f_j(x,t) = a_j \exp(2\kappa_j x - 8\kappa_j^3 t + \delta_j)$ for some constants a_j, κ_j and δ_j.

Taking $n = 1$, show that an *exact* solution for all ε is given by $f = 1 + \varepsilon f_1$, and thence if $a_1 = 1$ by

$$u(x,t) = -2\kappa_1^2\operatorname{sech}^2(\kappa_1 x - 4\kappa_1^3 t + \tfrac{1}{2}\delta_1 + \tfrac{1}{2}\ln\varepsilon)$$

for $\varepsilon > 0$. What solution u do you get for $\varepsilon < 0$?

Taking $n = 2$, show that an exact solution for all ε is given by

$$f = 1 + \varepsilon(f_1 + f_2) + \varepsilon^2\left(\frac{\kappa_2 - \kappa_1}{\kappa_2 + \kappa_1}\right)^2 f_1 f_2.$$

Verify that this solution can represent the interaction of two solitons.

[§7.3; Hirota (1971). The method can be used to derive an n-soliton solution by taking $f^{(1)} = \sum_{j=1}^{n} f_j$, it then being possible to show by mathematical induction that $f^{(m)} = 0$ identically for all $m > n$. It follows that $f = \det(f_{ij})$, where $f_{ij} = \delta_{ij} + 2\varepsilon\kappa_i f_i/(\kappa_i + \kappa_j)$. Hirota (1976) has generalized the transformation from u to f and thence found multisoliton solutions for several other nonlinear partial differential equations.]

4.15 *The structure of the argument that the eigenvalues of the bound states are constants.* Show that if $\psi(x,t)$ is a bound state with eigenvalue $\lambda < 0$ for a given well-behaved function $u(x,t)$, so that

$$(-D^2 + u)\psi = \lambda\psi,$$

where D denotes the differential operator d/dx, then

$$\lambda_t\psi = (-D^2 + u - \lambda)\psi_t + u_t\psi.$$

You may assume that u_t and u_{xxx} are continuous and that u vanishes sufficiently rapidly as $x \to \pm\infty$ for $t > 0$. Deduce that

$$\lambda_t\int_{-\infty}^{\infty}\psi^2 dx = \int_{-\infty}^{\infty}\{u_t\psi^2 + \psi_t(-D^2 + u - \lambda)\psi\}dx + [\psi_x\psi_t - \psi\psi_{xt}]_{-\infty}^{\infty}$$

$$= \int_{-\infty}^{\infty} u_t\psi^2 dx.$$

Hence or otherwise prove that if moreover $u_t = 6uu_x - u_{xxx}$ then

$$\lambda_t\int_{-\infty}^{\infty}\psi^2 dx = 0.$$

[§4.4. You are now ready to read Chapter 5.]

*CHAPTER 5 THE LAX METHOD

1 *Description of the method in terms of operators*

Why does the method of inverse scattering work? Is it just a revelation of a property possessed only by the KdV equation? In this chapter the success of the method will be explained by a deeper and more general argument due to Lax (1968). This abstract argument will show that the method may be applied to many, though not most, nonlinear initial-value problems.

The argument will use a little functional analysis, notably the ideas of inner product and symmetry in a Hilbert space. For further reading on applied functional analysis, the textbook by Griffel (1981) is recommended.

First, suppose that we wish to solve a nonlinear initial-value problem of the form

$$u_t = S(u), \tag{1}$$

where $u \in Y$ for each instant t, $S:Y \to Y$ is some nonlinear operator, and Y is some function space. Thus Y need not be a space of smooth scalar functions of x vanishing as $x \to \pm\infty$, and S need not be a partial differential operator, but equation (1) could be the KdV equation with $S(u) = 6uu_x - u_{xxx}$ for $-\infty < x < \infty$.

Next suppose that equation (1) can somehow be shown to be equivalent to the form

$$L_t = BL - LB, \tag{2}$$

where L and B are some linear spatial operators which depend upon u and thence t, and where L_t denotes the derivative of L with respect to the parameter t. We shall assume that L and B operate on elements of a Hilbert (i.e. complete inner-product) space H and also that L is symmetric. Even if an expression (2) of the original equation (1) exists, it may be difficult to find. We shall come back to this difficulty in the example below and in the problems. Indeed, finding the expression (2) is the crucial step of the solution of equation (1) by this method. But for the present we shall assume that expression (2) is known.

Note that the pair of operators L and B which satisfy equation (2) and imply equation (1) is certainly not unique; for the addition of any constant to B would imply that equations (1) and (2) are still satisfied.

Now we can consider the eigenvalue problem

$$L\psi = \lambda\psi \tag{3}$$

for all $t \geqslant 0$, where $\psi \in H$. Differentiating with respect to the parameter t, we see that

$$L\psi_t + L_t\psi = \lambda_t\psi + \lambda\psi_t$$

i.e., on use of equation (2),

$$\begin{aligned}\lambda_t\psi &= (L - \lambda)\psi_t + (BL - LB)\psi \\ &= (L - \lambda)\psi_t + B\lambda\psi - LB\psi \\ &= (L - \lambda)(\psi_t - B\psi).\end{aligned}$$

The inner product of ψ with this equation gives

$$\begin{aligned}\lambda_t(\psi,\psi) &= (\psi, (L - \lambda)(\psi_t - B\psi)) \\ &= ((L - \lambda)\psi, \psi_t - B\psi),\end{aligned}$$

because the operator L is symmetric,

$$= (0, \psi_t - B\psi)$$

$$= 0.$$

Therefore

$$\lambda_t = 0.$$

Therefore each eigenvalue of L is a constant. Thus the eigenvalues are invariant functionals of the solution $u(x,t)$ of equation (1).

It now follows that

$$(L - \lambda)(\psi_t - B\psi) = 0,$$

i.e. that $\psi_t - B\psi$ is an eigenfunction of L with eigenvalue λ. Therefore $\psi_t - B\psi \propto \psi$. So we may add the appropriate constant (or function of t) to B to ensure that

$$\psi_t = B\psi \qquad \text{for } t > 0. \tag{4}$$

What has this got to do with the method of inverse scattering? To answer the question, suppose that

$$L = - D^2 + u \quad \text{and} \quad B = - 4D^3 + 6uD + 3u_x + \alpha \tag{5}$$

where $D = \partial/\partial x$, α is independent of x, L and B operate on elements of H, and H is a space of square-integrable real functions. Therefore, by elementary calculus,

$$L\psi = - D^2\psi + u\psi,$$

$$BL\psi = (4D^5 - 4uD^3 - 12u_xD^2 - 12u_{xx}D - 4u_{xxx} - 6uD^3 + 6u^2D$$
$$+ 6uu_x - 3u_xD^2 + 3uu_x + \alpha L)\psi$$

and

$$LB\psi = (4D^5 - 6uD^3 - 12u_xD^2 - 6u_{xx}D - 3u_xD^2 - 6u_{xx}D - 3u_{xxx}$$
$$- 4uD^3 + 6u^2D + 3uu_x + \alpha L)\psi.$$

Therefore

$$(BL - LB)\psi = (-u_{xxx} + 6uu_x)\psi$$

for all sufficiently smooth $\psi \in H$. Therefore

$$BL - LB = -u_{xxx} + 6uu_x. \tag{6}$$

Also

$$L_t = u_t \tag{7}$$

simply. (Note that we differentiate the operator L with respect to its parameter t, not its operand ψ, so that $(L\psi)_t = L_t\psi + L\psi_t$.) Therefore $L_t = BL - LB$ if and only if $u_t = S(u)$, where the operator S is defined by

$$S(u) = -u_{xxx} + 6uu_x. \tag{8}$$

Thus the KdV equation is of the form (1) and its scattering problem is of the form (3). Also equation (4) becomes the evolution equation for ψ at the end of §4.4 on taking $\alpha = -4\kappa^3$ (so that ψ vanishes at infinity) and eliminating $\psi_{xxx} = \{(u - \lambda)\psi\}_x$.

Equation (2) is called a *Lax equation*. He was led to its formulation by recognizing that if L for all $t > 0$ is unitarily equivalent to L at $t = 0$ then it has constant eigenvalues and deducing that L satisfies an equation of the form (2) for an antisymmetric operator B (see Problem 5.5).

This is the fundamental background to why the eigenvalues of the scattering problem are constants as the solution of the KdV equation develops in time. It can be seen that we get an eigenvalue problem (3) for any nonlinear equation which can be put into the form (2) (see, for example, Problem 5.8) although deriving the form (2) from the form (1) is not always easy. Afterwards the scattering and inverse scattering theories appropriate to the eigenvalue problem (3) may be used to solve an initial-value problem for the nonlinear system (1). Chapter 4 may now be viewed as an example of this.

Problems

5.1 *Symmetry and antisymmetry of some operators.* Define L_2 as the Hilbert space of square-integrable real functions with inner product

$$(f,g) = \int_{-\infty}^{\infty} fg\,dx.$$

Further define the subspace $Y = \{f \,|\, f, f', f'', f''' \in L_2\}$ and operator $D = d/dx$. Deduce that, for a given function $u \in Y$, the operator $L = -D^2 + u$ is *symmetric* and the operator $B = -4D^3 + 6uD + 3(Du)$ is *antisymmetric*, i.e. $(f,Lg) = (Lf,g)$ and $(f,Bg) = -(Bf,g)\ \forall\, f,g \in Y$.

5.2 *An exercise with operators.* Show that if $L = -D^2 + u$ and $B = -cD$, where $D = \partial/\partial x$, $u = u(x,t)$ and c is a constant, then

$$BL - LB = -cu_x \quad \text{and} \quad L_t = u_t$$

identically. Hence show that the operator equation

$$L_t = BL - LB$$

implies that

$$u_t + cu_x = 0.$$

Find the general solution of this equation and confirm that the operator L of the Schrödinger problem has constant eigenvalues λ.

[Lax (1968, p.471).]

5.3 *A hierarchy of Lax equations.* Note that we have taken $L = -D^2 + u$ in Problem 5.2 with $B = B_0$ and in §5.1 with $B = B_1$, where

$$B_n = aD^{2n+1} + \sum_{m=1}^{n} \left(f_m D^{2m-1} + D^{2m-1} f_m \right)$$

for some constant a and functions f_m. Show that B_n is antisymmetric over an appropriate subspace of L_2.

Show that taking $B = aD^2 + fD + g$ for some constant a and functions f and g, and putting $L_t = BL - LB$ can give no new equation of the form $u_t = S(u)$.

Discuss the possibility of taking $B = B_n$ and L to solve a hierarchy of nonlinear partial differential equations of the form $u_t = S(u)$.

Discuss also the possibility of taking $L = D^4 + DuD + v$ to solve a pair of nonlinear equations of the form $u_t = P(u,v)$ and $v_t = Q(u,v)$.

[Lax(1968, p.471).]

5.4 *A derivation of the KdV equation.* Show that if there exist smooth functions $u(x,t)$ and non-null $\psi(x,t)$ such that $\psi_{xx} = (u + \kappa^2)\psi$ and $\psi_t = -4\psi_{xxx} + 6u\psi_x + 3u_x\psi - 4\kappa^3\psi$ for $\kappa > 0$ then $u_t = 6uu_x - u_{xxx}$.

[Hint: put $\psi_{xxt} = \psi_{txx}$ and eliminate higher derivatives of ψ.]

5.5 *Unitarily equivalent operators.* Let L be a linear operator which acts on a Hilbert space H and depends upon a parameter t. The *adjoint* U* of a linear operator U on H is defined so that the inner product $(f,Ug) = (U^*f,g)$ for all $f,g \in H$. Show that $L(t)$ has the same eigenvalues as $L(0)$ if $L(t)$ is *unitarily equivalent* to $L(0)$, i.e. if

$$U^*(t)L(t)U(t) = L(0) \quad \text{and} \quad U(t)U^*(t) = I,$$

where I is the identity operator.

Assuming that such an operator U does exist, show that there exists B on H such that

$$U_t = BU \quad \text{and} \quad B^* = -B.$$

Deduce that

$$L_t = BL - LB.$$

[Lax(1968).]

5.6 *A Lax equation for matrices.* Show that if $L(t)$ and $B(t)$ are $n \times n$ complex matrices such that

$$L_t = BL - LB,$$

L is Hermitian, and λ is a real eigenvalue of L, then λ is constant.

5.7 *An ordinary differential system.* Cast the system

$$\frac{dx}{dt} = gy, \quad \frac{dy}{dt} = -gx,$$

where $g = g(x,y,t)$ is some given continuous function, into the equivalent form,

$$L_t = BL - LB,$$

finding L as some symmetric and B as some antisymmetric real matrix whose elements depend upon x,y and g. Show that the eigenvalues of L are constants and hence deduce the (otherwise obvious) result that $x^2 + y^2$ is constant for each solution of the original differential system.

5.8 *The cubic Schrödinger equation.* Verify that if

$$L = i\begin{bmatrix} 1+p & 0 \\ 0 & 1-p \end{bmatrix}\frac{\partial}{\partial x} + \begin{bmatrix} 0 & \bar{u} \\ u & 0 \end{bmatrix},$$

$$B = ip\begin{bmatrix} 1 & 0 \\ 0 & 1 \end{bmatrix}\frac{\partial^2}{\partial x^2} + \begin{bmatrix} -i|u|^2/(1+p) & \bar{u}_x \\ -u_x & i|u|^2/(1-p) \end{bmatrix}$$

and

$$L_t = BL - LB,$$

where overbars denote complex conjugates, then u satisfies the cubic Schrödinger equation, namely

$$iu_t + u_{xx} + \nu|u|^2u = 0,$$

where $\nu = 2/(1 - p^2)$ and $0 < p < 1$.

[Zakharov & Shabat (1972, p.62).]

5.9 *The Kadomtsev-Petviashvili equation.* Show that if $L = -\partial^2/\partial x^2 + \partial/\partial y + u$, $B = -4\partial^3/\partial x^3 + 6u\partial/\partial x + 3u_x + 3\int^x u_y dx$ and $L_t = BL - LB$ then

$$(u_t - 6uu_x + u_{xxx})_x + 3u_{yy} = 0.$$

[Dryuma (1974).]

CHAPTER 6 THE SINE-GORDON EQUATION

1 *Introduction*

Klein (1927) and Gordon (1926) derived a relativistic wave equation of a charged particle in an electromagnetic field, using the recently discovered ideas of quantum theory. Their *Klein-Gordon equation* reduces to

$$\nabla^2\psi - \frac{1}{c^2}\frac{\partial^2\psi}{\partial t^2} = \left(\frac{mc}{h}\right)^2\psi \tag{1}$$

for the special case of a free particle. We may choose units so that $c = 1$ and generalize to consider the nonlinear form,

$$\frac{\partial^2\psi}{\partial t^2} - \nabla^2\psi + V'(\psi) = 0, \tag{2}$$

for some differentiable potential function V. In particular if $V'(\psi) = \sin\psi$ and $\partial/\partial y = \partial/\partial z = 0$, we get the *sine-Gordon equation*,

$$\psi_{tt} - \psi_{xx} + \sin\psi = 0. \tag{3}$$

The name was coined as a pun on Klein-Gordon.

We have introduced the sine-Gordon equation very briefly as a mathematical generalization. In fact it has a long history and many applications. A few are listed below.

(i) It seems to have arisen first at the end of the nineteenth century as an equation in the differential geometry of surfaces with constant Gaussian curvature $K = -1$ (cf. Eisenhart 1909, p.280).

(ii) It describes the propagation of a dislocation in a crystal whose periodicity is represented by $\sin\psi$ (Frenkel & Kontorova 1939).

(iii) It is a tentative model of an elementary particle (Perring & Skyrme 1962).

(iv) It is an equivalent form of the Thirring model (Coleman 1975).

(v) It describes the oscillations of rigid pendula attached to a stretched rubber band (Problem 6.2).

Before examining the solutions of equation (3), note that if we use characteristic coordinates

$$u = \tfrac{1}{2}(x - t) \quad \text{and} \quad v = \tfrac{1}{2}(x + t),$$

then it follows that $\psi_{xx} - \psi_{tt} = \psi_{uv}$ and therefore that

$$\psi_{uv} = \sin\psi. \tag{4}$$

The two forms, (3) and (4), of the sine-Gordon equation may be regarded as interchangeable. However, $t = 0$ corresponds to $u = v$, so the solution of initial-value problems depends on what variable is time.

Note also the invariance of the equation if $\psi \to \psi + 2n\pi$ for $n = \pm 1, \pm 2, \ldots$. Again, the transformation $\psi \to \psi + (2n + 1)\pi$ only leads to the replacement of $\sin\psi$ by $-\sin\psi$ in the equation.

If we linearize equation (3), then

$$\psi_{tt} - \psi_{xx} + \psi = 0. \tag{5}$$

Try normal modes with $\psi \propto e^{i(kx-\omega t)}$. This gives the dispersion relation,

$$\omega^2 = k^2 + 1. \tag{6}$$

Therefore ω is real for all real k, and the null solution $\psi = 0$ of (3) is stable. Similarly, if $\psi = \pi + \psi'$ for a small perturbation ψ', then we may linearize and find

$$\psi'_{tt} - \psi'_{xx} - \psi' = 0. \tag{7}$$

Taking $\psi' \propto e^{i(kx-\omega t)}$, we now get

$$\omega^2 = k^2 - 1. \tag{8}$$

Therefore $\omega^2 < 0$ if $0 \leq k^2 < 1$ and ψ' has a component growing exponentially in time, i.e. the equilibrium solution $\psi = \pi$ is unstable.

The stability and instability of these equilibrium solutions should have come as no surprise, for it may be easily seen that the sine-Gordon equation (3) becomes the equation for finite oscillations of a simple pendulum when the term in ψ_{xx} is neglected (and so when k is zero).

2 *Waves and solitons*

We first seek all possible waves of permanent form, using the methods of Chapter 2. Thus we seek solutions of the sine-Gordon equation (1.3) in the form $\psi(x,t) = f(X)$, where $X = x - Ut$. Then

$$U^2 f'' - f'' + \sin f = 0. \tag{1}$$

Therefore, on integration with respect to f,

$$\tfrac{1}{2}(U^2 - 1)f'^2 - \cos f = \text{constant}.$$

But $\cos f = 1 - 2\sin^2(\tfrac{1}{2}f)$. Therefore

$$\tfrac{1}{2}(U^2 - 1)f'^2 = A - 2\sin^2(\tfrac{1}{2}f) \equiv G(f), \text{ say,} \tag{2}$$

Fig. 1. Sketch of the graph of $G(f) = A - 2\sin^2(\tfrac{1}{2}f)$ for $0 < A < 2$.

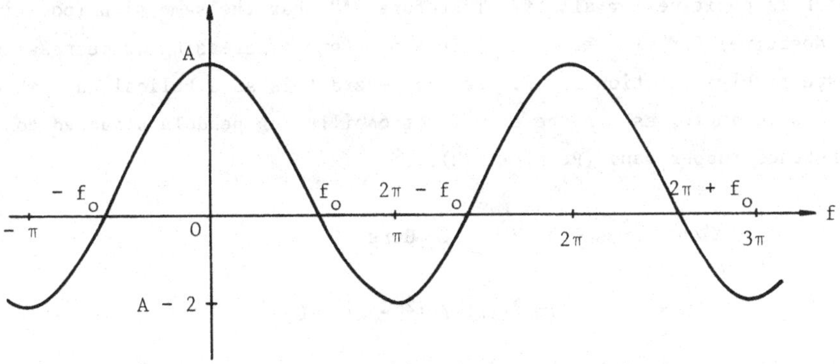

for some constant A of integration. Therefore

$$X = \int \frac{df}{\pm\sqrt{2\{A - 2\sin^2(\frac{1}{2}f)\}/(U^2 - 1)}} . \qquad (3)$$

Equation (2) shows that there are various cases according to the signs of $U^2 - 1$ and G, so that we must investigate the zeros of G by the method discussed in §2.3. (Remember that $f'^2 \geqslant 0$ in order that a real solution f exists.)

Case 1: $0 < A < 2$, $U^2 > 1$. In this case we can define $f_o = 2\sin^{-1}\sqrt{\frac{1}{2}A}$ with $0 < f_o < \pi$ so that f_o is the least positive zero of G. The graph of G in Fig. 1 shows where $G \geqslant 0$ and real solutions f are possible. In particular, we see that there are oscillations of the solution f, such that $-f_o \leqslant f \leqslant f_o$, with period

$$4\int_0^{f_o} \frac{df}{\sqrt{2\{A - 2\sin^2(\frac{1}{2}f)\}/(U^2 - 1)}} .$$

The solutions are analogues of the cnoidal waves of Chapter 2.

Case 2: $0 < A < 2$, $U^2 < 1$. Here we find periodic oscillations with $f_o - \pi \leqslant f - \pi \leqslant \pi - f_o$, because we require $G < 0$ when $U^2 < 1$ in order that f'^2 is positive. This case is similar to Case 1.

Case 3: $A < 0$, $U^2 < 1$. Here

$$f'^2 = 2\{2\sin^2(\tfrac{1}{2}f) - A\}/(1 - U^2),$$

which is positive for all f. Therefore f'' has the same sign (positive or negative) for all X, and f is a monotone (increasing or decreasing, respectively) function of X. We may regard this as a helical wave when ψ is an angle, as for the example of oscillating pendula attached to a stretched rubber band (Problem 6.2).

Case 4: $A > 2$, $U^2 > 1$. Here

$$f'^2 = 2\{A - 2\sin^2(\tfrac{1}{2}f)\}/(U^2 - 1) > 0,$$

and again we get a helical wave.

Case 5: $A = 0$, $U^2 < 1$. This is a limit of cases 2 and 3 above. Here equation (3) gives

$$X = \pm \sqrt{1 - U^2}\, \ell n\{\pm \tan(\tfrac{1}{4}f)\} + \text{constant}, \tag{4}$$

where the $\pm$ signs need not be related to one another. There are four subcases according to these signs.

Subcase (i). The solution

$$\tan(\tfrac{1}{4}f) = \exp\{(X - X_o)/\sqrt{1 - U^2}\}$$

gives f as increasing monotonically from 0 through π to 2π as X increases from $-\infty$ through X_o to $+\infty$.

Subcase (ii). The solution

$$\tan(\tfrac{1}{4}f) = -\exp\{-(X - X_o)/\sqrt{1 - U^2}\}$$

gives f as increasing monotonically from -2π through $-\pi$ to 0 as X increases from $-\infty$ through X_o to $+\infty$.

Subcase (iii). The solution

$$\tan(\tfrac{1}{4}f) = -\exp\{(X - X_o)/\sqrt{1 - U^2}\}$$

gives f as decreasing monotonically from 0 through $-\pi$ to -2π as X increases from $-\infty$ through X_o to $+\infty$.

Fig. 2. A sketch of the graph of a kink with $U^2 = \frac{3}{4}$: $f(X) = 4\tan^{-1}[\exp\{2(X - X_o)\}]$.

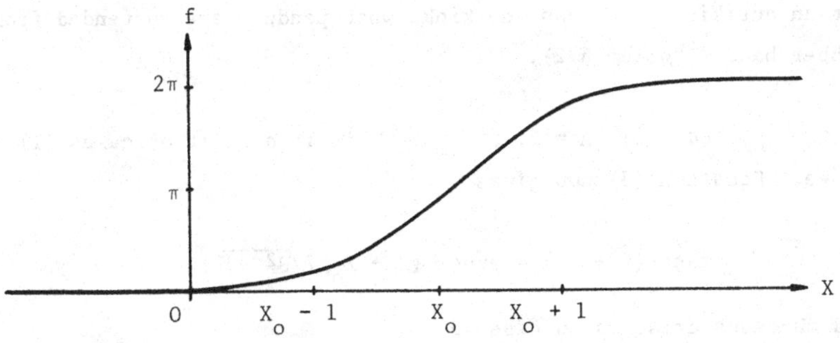

Subcase (iv). The last of the solutions,

$$\tan(\tfrac{1}{4}f) = \exp\{-(X - X_0)/\sqrt{1 - U^2}\},$$

gives f as decreasing monotonically from 2π through π to 0 as X increases from $-\infty$ through X_0 to $+\infty$.

Similar solutions arise when any integral multiple of 2π is added to f.

These are the soliton solutions of the sine-Gordon equation. The solutions $f(X)$ tend to constants as $X \to \pm\infty$, the constants differing by 2π. The solitons $\psi(x,t)$ propagate with velocity U as smooth step-like transitions between these constant levels, as shown in Fig. 2. The *derivatives* of ψ are hump-shaped, like a solitary wave. It can be seen that

$$\psi_x(x,t) = f'(x - Ut) \quad \text{and} \quad \psi_t(x,t) = -Uf'(x - Ut),$$

where equation (4) gives

$$f'(X) = \frac{\pm 4\tan(\tfrac{1}{4}f)}{\sqrt{1 - U^2}\,\{1 + \tan^2(\tfrac{1}{4}f)\}}$$

$$= \frac{\pm 2\operatorname{sech}\{(X - X_0)/\sqrt{1 - U^2}\}}{\sqrt{1 - U^2}} . \tag{5}$$

The width of the soliton has order of magnitude $\sqrt{1 - U^2}$, and the amplitude of f' is $\pm 2/\sqrt{1 - U^2}$.

A soliton for which f increases by 2π is sometimes called a *kink* and one for which f decreases by 2π an *antikink*; then f' represents a solitary wave of elevation for a kink, and one of depression for an antikink. You can see kinks when pendula are suspended from a rubber band (Problem 6.2).

Case 6: $A = 2$, $U^2 > 1$. This is a limit of cases (1) and (4) above. Equation (3) now gives

$$\tan\{\tfrac{1}{4}(f + \pi)\} = \exp\{\pm(X - X_0)/\sqrt{U^2 - 1}\},$$

and subcases arise as in case 5 .

In fact the solutions of cases 1, 2 and 6 are unstable and those of 3, 4 and 5 are stable. So the solutions of case 5 are those of special interest. We called them solitons, though we have yet to show that they retain their identities after interactions.

3 *Some other simple explicit solutions*

The form of the solution (2.4) suggests that the transformation

$$\phi = \tan(\tfrac{1}{4}\psi) \tag{1}$$

may be of further use. Then the sine-Gordon equation (1.3) becomes

$$(1 + \phi^2)(\phi_{tt} - \phi_{xx} + \phi) - 2\phi(\phi_t^2 - \phi_x^2 + \phi^2) = 0, \tag{2}$$

on use of the trignometric identity $\sin\psi = 4\phi(1 - \phi^2)/(1 + \phi^2)^2$. We can seek to separate the variables by trial of

$$\phi(x,t) = f(x)/g(t) \tag{3}$$

for some functions f and g to be found, noting that the solution (2.4) for a soliton is a special case of this form. Substitution into equation (2) now gives

$$(f^2 + g^2)\left(\frac{f''}{f} + \frac{\ddot{g}}{g}\right) + f^2 - g^2 = 2(f'^2 + \dot{g}^2). \tag{4}$$

where $f' = df/dx$ and $\dot{g} = dg/dt$ etc. Differentiating this equation with respect to x and t, we find

$$(g^2)^{\cdot}\left(\frac{f''}{f}\right)' + (f^2)'\left(\frac{\ddot{g}}{g}\right)^{\cdot} = 0.$$

for all x and t. Therefore

$$\frac{1}{(f^2)'}\left(\frac{f''}{f}\right)' = -\frac{1}{(g^2)^{\cdot}}\left(\frac{\ddot{g}}{g}\right)^{\cdot} = 2\mu, \tag{5}$$

say, where μ is a separation constant. Each of the two ordinary differential equations (5) may be integrated twice to give

$$\tfrac{1}{2}f'^2 = \tfrac{1}{2}\mu f^4 + \tfrac{1}{2}c_1 f^2 + c_2 \quad \text{and} \quad \tfrac{1}{2}\dot{g}^2 = -\tfrac{1}{2}\mu g^4 + \tfrac{1}{2}d_1 g^2 + d_2,$$

where c_1, c_2, d_1 and d_2 are integration constants. Substitution of these equations into (4) shows that $c_1 - d_1 = 1$ and $c_2 + d_2 = 0$. Therefore

$$f'^2 = \mu f^4 + (1 + \lambda)f^2 + \nu \quad \text{and} \quad \dot{g}^2 = -\mu g^4 + \lambda g^2 - \nu, \tag{6}$$

for some constants λ, μ and ν.

Equations (6) can be solved generally in simple terms of elliptic functions, but they also have some special solutions in terms of elementary functions in addition to those we have already met for the special case when $\mu = \nu = 0$.

4 *The interaction of two solitons*

There are also elementary solutions of equations (3.6) if $\lambda > 0$ and $\mu = 0$. Then

$$f'^2 = (1 + \lambda)f^2 + \nu \quad \text{and} \quad \dot{g}^2 = \lambda g^2 - \nu \tag{1}$$

so that

$$f(x) = \pm\sqrt{\frac{\nu}{1 + \lambda}}\,\sinh\{\sqrt{1 + \lambda}(x - x_0)\} \quad \text{and}$$

$$g(t) = \pm\sqrt{\frac{\nu}{\lambda}}\,\cosh\{\sqrt{\lambda}(t - t_0)\} \tag{2}$$

if $\nu > 0$, where x_0 and t_0 are constants of integration. (If $\nu < 0$, then the cosh and sinh are interchanged and ν is replaced by $-\nu$ in solution (2). See Problem 6.8.) We may put $x_0 = 0$ and $t_0 = 0$ without loss of generality. Now equations (3.1) and (3.3) give, independently of the value of ν,

$$\phi = \tan(\tfrac{1}{4}\psi) = \frac{U\sinh(x/\sqrt{1 - U^2})}{\cosh(Ut/\sqrt{1 - U^2})}, \tag{3}$$

say, where $\lambda = U^2/(1 - U^2)$ so $0 < U^2 < 1$.

This solution was first found by Perring & Skyrme (1962) very differently. It seems that they guessed it as a fit to some numerical calculations they made of the interaction of two solitons and then verified it analytically. To show that the solution does indeed represent the interaction of two equal solitons, we evaluate it asymptotically as $t \to \pm\infty$. First take $U > 0$, without loss of generality. It follows at once that

Fig. 3. Sketches of a solution representing the interaction of two equal solitons:
$\psi(x,t) = 4\tan^{-1}\{U\sinh(x/\sqrt{1-U^2})/\cosh(Ut/\sqrt{1-U^2})\}$ for $U > 0$.
(a) $t \ll -1$. (b) $t = 0$. (c) $t \gg 1$.

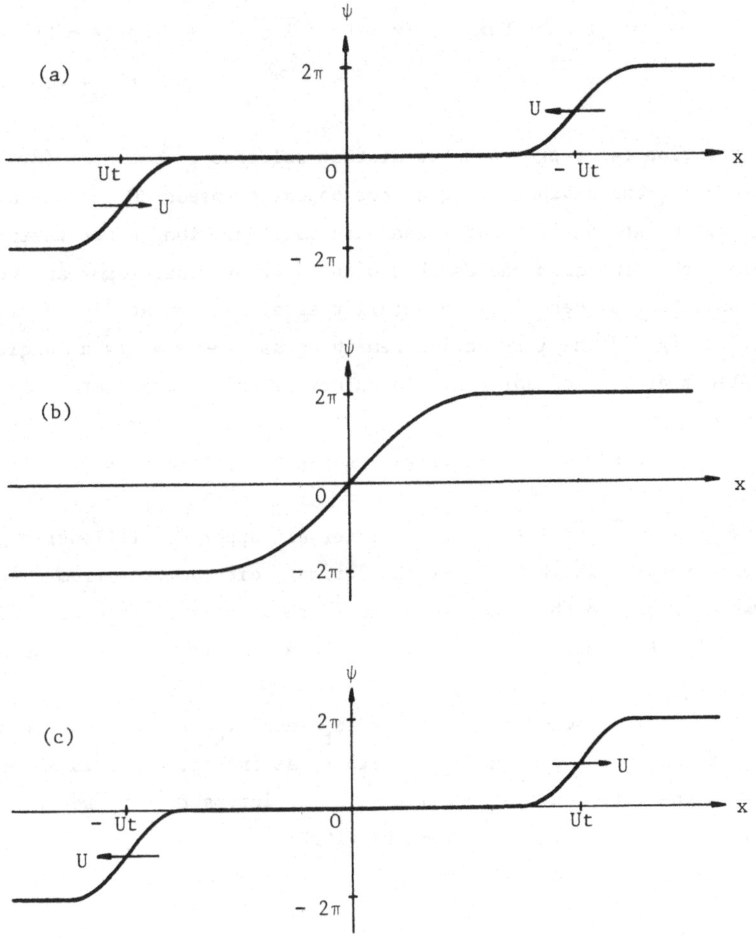

$$\phi(x,t) \sim U \exp\{(x + Ut)/\sqrt{1 - U^2}\} - U \exp\{-(x - Ut)/\sqrt{1 - U^2}\} \tag{4}$$

uniformly for all x as $t \to -\infty$, because

$$\cosh(Ut/\sqrt{1 - U^2}) \sim \tfrac{1}{2}\exp(U|t|/\sqrt{1 - U^2})$$

and

$$\sinh(x/\sqrt{1 - U^2}) = \tfrac{1}{2}\exp(x/\sqrt{1 - U^2}) - \tfrac{1}{2}\exp(-x/\sqrt{1 - U^2}).$$

Similarly

$$\phi(x,t) \sim -U\exp\{-(x + Ut)/\sqrt{1 - U^2}\} + U\exp\{(x - Ut)/\sqrt{1 - U^2}\} \quad \text{as } t \to +\infty. \tag{5}$$

Now relation (4) represents two distant solitons of the form (2.4) approaching one another at equal but opposite speeds U, as shown in Fig. 3. Relation (5) shows these same solitons, long after their interaction, receding from one another with their original forms and velocities. The solution (3) describes in detail the interaction at finite times. The only vestige of the interaction remaining as $t \to +\infty$ is a longitudinal displacement of each soliton. To calculate this, note that

$$\pm U \exp\{\pm (x + Ut)/\sqrt{1 - U^2}\} = \pm \exp\{\pm (x + Ut \pm \tfrac{1}{2}\delta)/\sqrt{1 - U^2}\},$$

where $\delta = 2\sqrt{1 - U^2}\,\ell n(1/U)$ and either all upper or all lower signs are taken; this shows that δ is the ultimate displacement from what the first soliton's path would have been if there were no interaction. Similarly, the second soliton can be shown to have the same displacement δ.

This solution generates ψ_t and ψ_x, whose graphs represent the interaction of hump-shaped solitons, as in Fig. 4. This shows more clearly the collison of two solitons. The interaction is better seen as a motion picture (Eilbeck & Lomdahl F1982).

Fig. 4 Sketches of the graphs of

$$\psi_x = \frac{4U\cosh(x/\sqrt{1-U^2})\cosh(Ut/\sqrt{1-U^2})}{\sqrt{1-U^2}\{\sinh^2(x/\sqrt{1-U^2}) + \cosh^2(Ut/\sqrt{1-U^2})\}} \quad \text{for} \quad U > 0.$$

representing the interaction of two solitons.
(a) $t << -1$. (b) $t = 0$. (c) $t >> 1$.

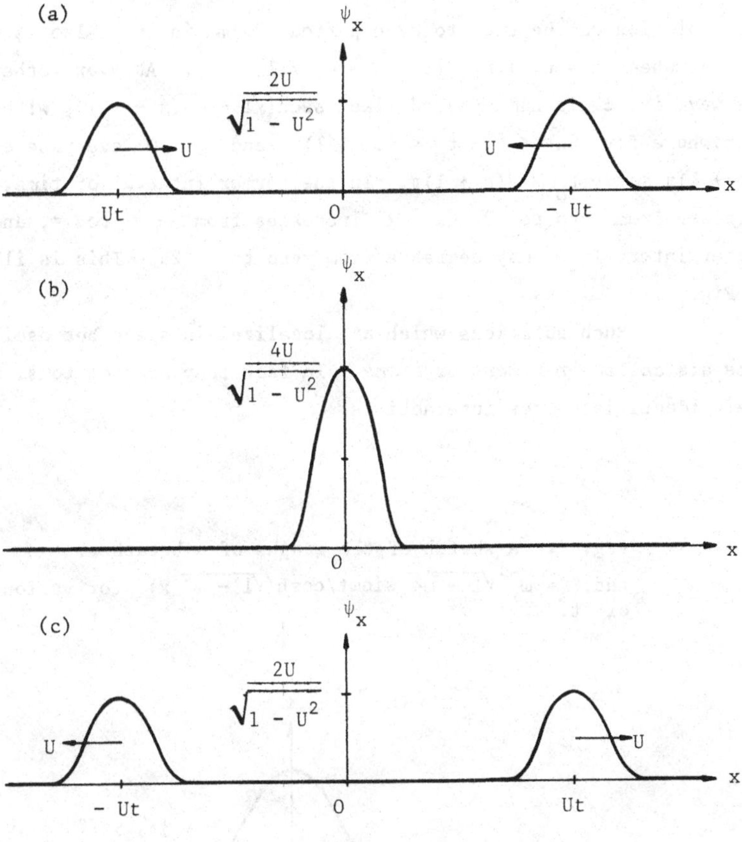

5 *A breather*

If $\lambda < 0$ and $\nu > 0$, then we may put $\lambda = -\omega^2$, translate ψ by 2π, and replace the solution (4.3) by

$$\phi = \tan(\tfrac{1}{4}\psi) = \frac{\sqrt{1-\omega^2}}{\omega}\,\frac{\sin\{\omega(t-t_0)\}}{\cosh\{\sqrt{1-\omega^2}(x-x_0)\}}\,. \tag{1}$$

This solution can be seen to have period $2\pi/\omega$ in t. Also $\phi = 0$ for all x when $t = t_0 + n\pi$ for $n = 0, \pm 1, \ldots$. At every other instant the wave in ϕ is hump-shaped like $\operatorname{sech}\{\sqrt{1-\omega^2}(x-x_0)\}$, with $\phi > 0$ everywhere for $2n\pi < t - t_0 < (2n+1)\pi$ and $\phi < 0$ everywhere for $(2n+1)\pi < t - t_0 < 2(n+1)\pi$. In the former interval of time, ψ may increase from zero to 2π as x increases from $-\infty$ to ∞, and in the latter interval ψ may decrease from zero to -2π. This is illustrated in Fig. 5.

Such solutions which are localized in space but oscillate in time are called *breathers* or *bions*. In fact they are solitons, retaining their identities after interactions.

Fig. 5. A sketch of the graphs of a breather: $\tan\frac{1}{4}\psi = \omega^{-1}\sqrt{1-\omega^2}\,\sin\omega t/\cosh(\sqrt{1-\omega^2}\,x)$ for various values of t.

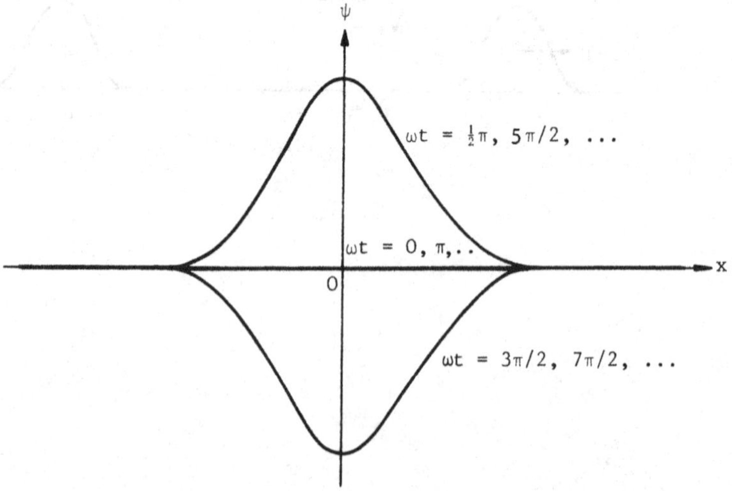

6 *The method of inverse scattering*

A method of inverse scattering may be used to show that kinks and breathers are indeed solitons. Ablowitz *et al.* (1973) devised yet another ingenious method to solve initial-value problems, taking the sine-Gordon equation in the characteristic form,

$$\frac{\partial^2 \psi}{\partial x \partial t} = \sin\psi. \tag{1}$$

(Remember, however, that the initial-value problem for equation (1.3) is not quite the same.) Consider, then, the complex linear system of equations,

$$\frac{\partial v_1}{\partial x} + i\zeta v_1 = q v_2 \tag{2}$$

and

$$\frac{\partial v_2}{\partial x} - i\zeta v_2 = -\bar{q} v_1, \tag{3}$$

which, together with the conditions that v_1 and v_2 are bounded as $x \to \pm\infty$, may be regarded as a problem to determine the eigenvalues ζ for given function q.

Even though this eigenvalue problem is not symmetric, we may follow the spirit of Chapter 4. First note that if v_1 and v_2 are bounded and q tends to zero sufficiently rapidly as $x \to \pm\infty$ then $v_1 \sim \text{constant} \times e^{-i\zeta x}$ and $v_2 \sim \text{constant} \times e^{i\zeta x}$ as $x \to \pm\infty$. Therefore there exists an unbound state if and only if ζ is real, as is shown in Problem 6.11. It follows that ζ may be taken to be any fixed real number and so that $\zeta_t = 0$. Next we may deduce that if $q = -\frac{1}{2}\psi_x$ and ψ satisfies the sine-Gordon equation (1) then

$$\frac{\partial v_1}{\partial t} = \frac{i}{4\zeta}(v_1\cos\psi + v_2\sin\psi) \tag{4}$$

and

$$\frac{\partial v_2}{\partial t} = \frac{i}{4\zeta}(v_1\sin\psi - v_2\cos\psi), \tag{5}$$

by differentiating equations (2) and (3) with respect to t and integrating with respect to x.

Alternatively, we may assume that v_1 and v_2 evolve according to equations (4) and (5) and verify that this is consistent with equations (1), (2) and (3) for a constant eigenvalue ζ. To do this, note that $\partial(2)/\partial t - (\partial/\partial x + i\zeta)(4)$ gives

$$i\zeta_t v_1 = 0$$

on elimination of v_{2t} from equation (5), v_{1x} from (2) and v_{2x} from (3), if ψ satisfies (1). A similar operation with $\partial(3)/\partial t - (\partial/\partial x - i\zeta)(5)$ confirms the result that $\zeta_t = 0$.

Equations (2) and (3) are not quite equivalent to the Schrödinger eigenvalue problem, but their solutions are similar. The method of inverse scattering may be applied similarly to solve initial-value problems for the sine-Gordon equation (1): (i) solve the direct scattering problem with the initial distribution $q(x,0)$ to find the scattering data at $t = 0$; (ii) use the evolution equations to find the scattering data for $t > 0$; (iii) invert the scattering problem, finding q for $t > 0$ from the solution of a linear integral equation; and (iv) integrate q to find ψ for $t > 0$.

The principles of the method should be familiar to those who understand Chapter 4. The details are given by Zakharov & Shabat (1972), who inverted a similar scattering problem to solve the cubic Schrödinger equation (see Problem 5.8) and by Ablowitz *et al.* (1973), who devised the method above. This method can be used to solve other nonlinear equations, e.g. the modified KdV equation and the cubic Schrödinger equation by changing the details of the scattering equations (2) and (3) (Ablowitz *et al.* 1974). For it can be seen from Problem 5.8 that the scattering problem for the cubic Schrödinger equation may be cast into the form of equations (2) and (3) by taking $q = iu/\sqrt{1 - p^2}$ and $\zeta = \lambda p/(1 - p^2)$. In this introductory text we shall not explain the details of the method further. However, the reader should recognize the power of the method, and follow up the references if he wishes.

Problems

6.1 *Motion pictures of soliton interactions.* See the animated film of solitons by Eilbeck & Lomdahl (F1982), and relate what you see to the theoretical results you have learnt.

6.2 *A simple experimental model for the sine-Gordon equation.* Obtain a length (100-150cm, say) of rubber band, about 6mm wide and 2mm thick. Insert carefully a set of similar pins with large heads into one side of the strip, placing them uniformly about 3mm apart. Clamp each end of the strip so that the pins hang downwards as a row of simple pendula. See Fig. 6.

Using the equation

$$\ell \frac{d^2\psi}{dt^2} = - g \sin\psi$$

for the motion of a simple pendulum of length ℓ which makes the finite angle ψ with the vertical, and the equation

$$\frac{\partial^2\psi}{\partial x^2} = c^2 \frac{\partial^2\psi}{\partial t^2}$$

for torsional waves along an elastic column, argue plausibly that the angle of your pins is governed by a sine-Gordon equation.

Seek to develop kinks and kink interactions with your pins and rubber strip (but do not let your expectations of too good an experimental model of the sine-Gordon equation bring you disappointment).

[Scott (1970, §§2.7, 5.5).]

Fig. 6 A sketch of a short stretch of the rubber band.

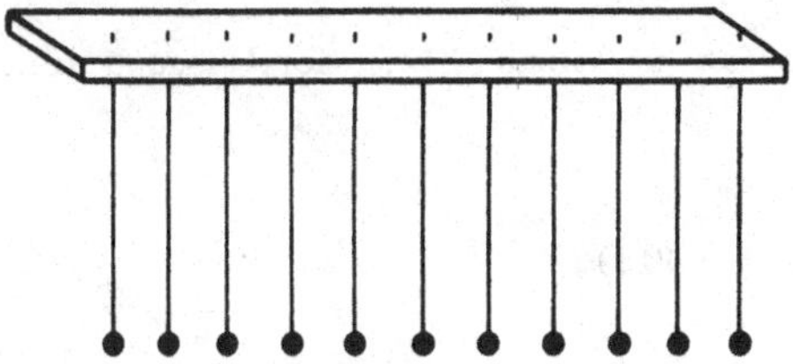

6.3 *Periodic solutions of nonlinear Klein-Gordon equations.* Seek solutions of

$$\psi_{tt} - \psi_{xx} + V'(\psi) = 0$$

which represent waves of permanent form by trying $\psi(x,t) = f(\theta)$, where $\theta = kx - \omega t$. Deduce that

$$\theta = \sqrt{\tfrac{1}{2}(\omega^2 - k^2)} \int \frac{df}{\pm\sqrt{E - V(f)}},$$

where E is a constant of integration. Show that if k is chosen (as it may be without loss of generality) so that f has period 2π then

$$2\pi = \sqrt{\tfrac{1}{2}(\omega^2 - k^2)} \oint \frac{df}{\pm\sqrt{E - V(f)}},$$

where the integration is over a complete period, from one simple zero of $E - V$ to the next one and back again, on taking the signs $\pm$ appropriately.

If $V(\psi) = \tfrac{1}{2}\psi^2 + b\psi^4 + o(\psi^4)$ as $\psi \to 0$, deduce that

$$f = a\cos\theta + \frac{1}{8} a^3 b \cos 3\theta + o(a^3) \quad \text{as } a \to 0,$$

where the dispersion relation is given by

$$\omega^2 = k^2 + 1 + 3a^2 b + o(a^2) \quad \text{as } a \to 0$$

and where

$$E = \tfrac{1}{2}a^2 + \frac{9}{8} a^4 b + o(a^4).$$

[Whitham (1974, §14.1).]

6.4 *A variational principle for nonlinear Klein-Gordon equations.* Using the Lagrangian density

$$\mathscr{L} = \tfrac{1}{2}\{\psi_t^2 - (\nabla\psi)^2\} - V(\psi),$$

deduce the nonlinear Klein-Gordon equation (1.2) by requiring that the integral $\int \mathscr{L} d\underset{\sim}{x} dt$ is stationary with respect to small variations of ψ.

Find the Hamiltonian $\mathscr{H} = \psi_t \partial\mathscr{L}/\partial\psi_t - \mathscr{L}$. Deduce that the total 'energy' $\int \mathscr{H}\, d\underset{\sim}{x} = \int [\frac{1}{2}\{\psi_t^2 + (\nabla\psi)^2\} + V(\psi)]d\underset{\sim}{x}$.

6.5 *Conservation laws for the sine-Gordon equation.* Given the sine-Gordon equation in the characteristic form

$$\psi_{uv} = \sin\psi,$$

verify the following conservation relations:

$$(\tfrac{1}{2}\psi_v^2)_u - (1 - \cos\psi)_v = 0,$$

$$(1 - \cos\psi)_u - (\tfrac{1}{2}\psi_u^2)_v = 0,$$

$$(\tfrac{1}{4}\psi_v^4 - \psi_{vv}^2)_u + (\psi_v^2\cos\psi)_v = 0$$

$$\left(\frac{1}{6}\psi_v^6 - \frac{2}{3}\psi_v^2\psi_{vv}^2 + \frac{8}{9}\psi_v^3\psi_{vvv} + \frac{4}{3}\psi_{vvv}^2\right)_u + \left\{\left(\frac{1}{9}\psi_v^4 - \frac{4}{3}\psi_{vv}^2\right)\cos\psi\right\}_v = 0.$$

[Lamb (1970).]

6.6 *Similarity solution.* Show that the sine-Gordon equation,

$$\psi_{uv} = \sin\psi,$$

is invariant under the continuous group $u \to ku$ and $v \to v/k$.

Seeking invariant solutions of the form $\psi = f(X)$, where $X = uv$, show that

$$Xf'' + f' = \sin f.$$

Hence show that $w = e^{if}$ satisfies

$$w'' - w'^2/w + (2w' - w^2 + 1)/2X = 0.$$

The solutions of this equation are a special case of what are called the third *Painlevé transcendents* (Ince 1927, p.345).

6.7 *Verification of a soliton solution.* Show that when $\lambda > 0$ and $\mu = \nu = 0$ equations (3.6) have a general solution of the form

$$f(x) = \pm \exp\{\pm (x - x_0)/\sqrt{1 - U^2}\}$$

and

$$g(t) = \exp\{\pm U(t - t_0)/\sqrt{1 - U^2}\},$$

where $U^2/(1 - U^2) = \lambda$, so that equations (3.1) and (3.3) give

$$\tan(\tfrac{1}{4}\psi) = \pm \exp[\{\pm (x - x_0) \pm U(t - t_0)\}/\sqrt{1 - U^2}],$$

in agreement with equation (2.4).

6.8 *The interaction of a soliton and an antisoliton.* Verify that

$$\phi = \tan(\tfrac{1}{4}\psi) = \frac{U\cosh(x/\sqrt{1 - U^2})}{\sinh(Ut/\sqrt{1 - U^2})}$$

is an exact solution of the sine-Gordon equation (1.3). Interpret the solution as the interaction of a kink and an antikink by examining the asymptotic behaviour of ψ as $t \to \pm \infty$. Sketch the solutions, noting that ψ is an even function of x, is an odd function of t, and is instantaneously a constant at $t = 0$.

6.9 *Small-amplitude breather.* Approximate the solution (5.1), namely

$$\tan(\tfrac{1}{4}\psi) = \frac{\sqrt{1 - \omega^2}}{\omega} \frac{\sin\{\omega(t - t_0)\}}{\cosh\{\sqrt{1 - \omega^2}(x - x_0)\}},$$

where $\sqrt{1 - \omega^2} = \varepsilon \ll 1$, showing that

$$\psi \sim -4\varepsilon \operatorname{sech}\{\varepsilon(x - x_0)\}\operatorname{Re}[i \exp\{i(1 - \tfrac{1}{2}\varepsilon^2)(t - t_0)\}]$$

as $\varepsilon \to 0$.

6.10 *Second-order equation for scattering problem.* Show that if

$$v_{1x} + i\zeta v_1 = qv_2, \quad v_{2x} - i\zeta v_2 = -\bar{q}v_1$$

and $q \neq 0$, then

$$q(v_{1x}/q)_x + (|q|^2 + \zeta^2 - i\zeta q_x/q)v_1 = 0.$$

Show further that if $w = v_1 \pm iv_2$ and $\bar{q} = q$ then

$$w_{xx} + (|q|^2 + \zeta^2 \pm iq_x)w = 0.$$

6.11 *The scattering problem.* Define $\underset{\sim}{u}(x,\zeta,t) = \begin{bmatrix} u_1 \\ u_2 \end{bmatrix}$ as the solution of equations (6.2) and (6.3) such that

$$\underset{\sim}{u} \sim \begin{bmatrix} 1 \\ 0 \end{bmatrix} e^{-i\zeta x} \qquad \text{as} \quad x \to -\infty.$$

Similarly let $\underset{\sim}{v}(x,\zeta,t)$ be the solution such that

$$\underset{\sim}{v} \sim \begin{bmatrix} 0 \\ 1 \end{bmatrix} e^{i\zeta x} \qquad \text{as} \quad x \to +\infty.$$

First show that $\underset{\sim}{v}^*(x,\zeta,t) = \begin{bmatrix} \bar{v}_2(x,\bar{\zeta},t) \\ -\bar{v}_1(x,\bar{\zeta},t) \end{bmatrix}$ is a solution of equations (6.2) and (6.3) with ζ replaced by $\bar{\zeta}$, where overbars denote complex conjugates. Deduce that $\underset{\sim}{v}^*$ is independent of $\underset{\sim}{v}$ and therefore that

$$\underset{\sim}{u} = a\underset{\sim}{v}^* + b\underset{\sim}{v}$$

when ζ is real, for some complex coefficients $a(\zeta,t)$ and $b(\zeta,t)$.

Next show that if $\underset{\sim}{y}$ and $\underset{\sim}{z}$ are any pair of solutions of equations (6.2) and (6.3) then the derivative of their Wronskian is zero, i.e.

$$\frac{d}{dx}(y_1 z_2 - y_2 z_1) = 0.$$

Choosing $\underset{\sim}{y} = \underset{\sim}{u}$ and $\underset{\sim}{z} = -\underset{\sim}{u}^*$, deduce that if ζ is real then

$$|a|^2 + |b|^2 = 1$$

and

$$a = u_1 v_2 - u_2 v_1 .$$

Hence show by analytic continuation that

$$\underset{\sim}{u}(x,\zeta_n,t) = c_n \underset{\sim}{v}(x,\zeta_n,t) \qquad \text{for} \quad n = 1,2,\ldots,p$$

and for some coefficients $c_n(t)$, where p is some non-negative integer and the zeros ζ_n of $a(\zeta,t)$ in the upper half of the complex plane are the discrete eigenvalues of the system.

*Show further, from the evolution equation for $\underset{\sim}{u}$, that

$$a(\zeta,t) = a(\zeta,0), \quad b(\zeta,t) = b(\zeta,0)e^{-it/2\zeta}$$

and

$$c_n(t) = c_n(0)e^{-it/2\zeta} \qquad \text{for} \quad t > 0.$$

[Zakharov & Shabat (1972); Ablowitz *et al.* (1973).]

6.12 *The reduced Maxwell-Bloch equations.* In the theory of nonlinear optics, it has been found that an ultrashort pulse of light in a dielectric medium of two-level atoms is governed by the system of equations,

$$E_t = s, \quad s_x = Eu + \mu r, \quad r_x = -\mu s \quad \text{and} \quad u_x = -Es,$$

where E,s,u and r are functions of x and t, and μ is a constant.

Show that when $\mu = 0$ this system has solutions such that $u = \cos\phi$, where $E = \phi_x$, and hence deduce that ϕ satisfies a sine-Gordon equation.

Discuss the wave solutions of permanent form when $\mu \neq 0$. In particular, seeking solitary waves for which $E,r,s \to 0$ and $u \to -1$ as $x \to \pm\infty$, show that

$$E(x,t) = a\ \mathrm{sech}[\tfrac{1}{2}a\{x - 4t/(4\mu^2 + a^2) + \delta\}]$$

gives a solution for all a and δ.

[Lamb (1971); Gibbon *et al.* (1973) used the method of inverse scattering to solve the initial-value problem for this system and to generate multisoliton solutions.]

6.13 *The scattering and evolution equations in terms of operators.* Given that

$$V_x = i\zeta MV + NV, \quad V_t = BV$$

and $\zeta_t = 0$ for all vectors V belonging to some space, parameters ζ, and linear operators M, N and B, use the formula $V_{tx} = V_{xt}$ to show that

$$N_t + i\zeta M_t = B_x + BN - NB + i\zeta(BM - MB).$$

Taking $B = \dfrac{i}{4\zeta}\begin{bmatrix}\cos\psi & \sin\psi \\ \sin\psi & -\cos\psi\end{bmatrix}$, $M = \begin{bmatrix}-1 & 0 \\ 0 & 1\end{bmatrix}$ and $N = \tfrac{1}{2}\psi_x\begin{bmatrix}0 & -1 \\ 1 & 0\end{bmatrix}$, deduce that

$$\psi_{xt} = \sin\psi.$$

CHAPTER 7 BÄCKLUND TRANSFORMATIONS

1 *Introduction*

Bäcklund transformations were devised in the 1880s for use in the theories of differential geometry and of differential equations (cf. Forsyth (1906, p.433) and Eisenhart (1909, p.284)). They arose as a generalization of contact transformations. Perhaps the simplest example of a Bäcklund transformation is the Cauchy-Riemann relations (see Problem 7.1). You may also have glimpsed what the transformations are if you have studied the solution of the Burgers equation (Problem 4.2) and Miura's transformation (Problem 4.3). Indeed, a Bäcklund transformation seems to be associated with the applicability of the method of inverse scattering itself (note Problem 5.4). However, here we shall treat the transformation merely as a topic in advanced calculus in order to generate solutions representing interactions of solitons.

A Bäcklund transformation is essentially defined as a pair of partial differential relations involving two independent variables and their derivatives which together imply that each one of the dependent variables satisfies separately a partial differential equation. Thus, for example, the transformation

$$v_x = F(u,v,u_x,u_y,x,y) \quad \text{and} \quad v_y = G(u,v,u_x,u_y,x,y) \tag{1}$$

would imply that u and v satisfy partial differential equations of the operational form,

$$P(u) = 0 \quad \text{and} \quad Q(v) = 0. \tag{2}$$

To illustrate this first, it seems best to use a simple example which has, however, no direct connection with solitons. So we consider *Liouville's equation* in characteristic form, namely

$$u_{xy} = e^u. \tag{3}$$

We shall solve this equation by introducing a new dependent variable v and the following pair of first-order equations,

$$u_x + v_x = \sqrt{2}\, e^{\frac{1}{2}(u-v)} \tag{4}$$

and

$$u_y - v_y = \sqrt{2}\, e^{\frac{1}{2}(u+v)}. \tag{5}$$

These are an example of a Bäcklund transformation from u to v (or from v to u). Now differentiate equation (4) with respect to y and use equation (5) to find that

$$u_{xy} + v_{xy} = e^u.$$

Also, cross-differentiating to form $\partial(4)/\partial y - \partial(5)/\partial x$, and using (4) and (5), we find that

$$v_{xy} = 0. \tag{6}$$

Thus equations (4) and (5) imply (3) and (6).

The advantage of all this comes from being able to solve equation (6) easily and hence generate all solutions of equation (3) by integrating the Bäcklund transformation. The general solution of equation (6) is, of course,

$$v(x,y) = \phi(x) + \psi(y) \tag{7}$$

for arbitrary differentiable functions ϕ and ψ. Then equation (4) becomes

$$u_x + \phi' = \sqrt{2}\, e^{\frac{1}{2}(u-\phi-\psi)}.$$

Therefore

$$e^{\phi}(u + \phi - \psi)_x = \sqrt{2}\, e^{\frac{1}{2}(u+\phi-\psi)}.$$

Now define a new independent variable X such that

$$X(x) = \int^x e^{-\phi}.$$

Therefore

$$(u + \phi - \psi)_X = \sqrt{2}\, e^{\frac{1}{2}(u+\phi-\psi)}.$$

Therefore

$$-\sqrt{2}\, e^{-\frac{1}{2}(u+\phi-\psi)} = X + g(y),$$

where g is an arbitrary function of integration. Similarly, equation (5) gives

$$-\sqrt{2}\, e^{-\frac{1}{2}(u+\phi-\psi)} = Y + h(x),$$

where

$$Y(y) = \int^y e^{\psi}.$$

Therefore

$$e^{\frac{1}{2}(u+\phi-\psi)} = -\sqrt{2}/(X + Y).$$

Therefore

$$u = \psi - \phi + 2\ell n\{-\sqrt{2}/(X + Y)\}. \tag{8}$$

This is the general solution of equation (3) which Liouville (1853) first found by a different method.

So we shall regard a Bäcklund transformation as a pair of partial differential relations (for example, (4) and (5)) which determine a solution of a given partial differential equation of higher order (for example, the solution u of (3)) in terms of a solution of another equation (for example, the solution v of (6)) or another solution of the same equation. We also regard equation (6) as a consistency condition in order that there exists a solution u of equations (4) and (5) for given v. More general treatments than the present one are given by Forsyth (1906, Chap. XXI) and Eisenhart (1909, pp.280-91). These books give some motivation for the selection of a Bäcklund transformation. In particular, Clairin (1903, see Forsyth (1906, p.433)) devised a technique to construct and classify Bäcklund transformations. However, Bäcklund transformations cannot be found for all nonlinear partial differential equations.

2 *The sine-Gordon equation*

Take the sine-Gordon equation in its characteristic form,

$$\psi_{uv} = \sin\psi. \tag{1}$$

Trial and error might lead to the Bäcklund transformation from ψ to ϕ,

$$\phi_u = \psi_u + 2\lambda\sin\{\tfrac{1}{2}(\phi + \psi)\} \tag{2}$$

and

$$\phi_v = -\psi_v + 2\lambda^{-1}\sin\{\tfrac{1}{2}(\phi - \psi)\}, \tag{3}$$

where λ is an arbitrary real parameter. It seems (Forsyth 1906, p.454; Eisenhart 1909, p.281) that this choice of transformation is due to Bianchi. Assuming the transformation, we form $\partial(2)/\partial v + \partial(3)/\partial u$ and use equations (2) and (3) to deduce that

$$\phi_{uv} = \sin\phi. \tag{4}$$

Similarly, $\partial(2)/\partial v - \partial(3)/\partial u$ gives equation (1). Thus equations (2) and (3) imply equations (1) and (4). In this example equation (4) happens to be the same as (1).

Next we shall use the transformation. Consider the trivial solution $\psi = \psi_0$ of the sine-Gordon equation (1) where $\psi_0 = 0$ is the null function. Now if $\psi = 0$ in the Bäcklund transformation, (2) and (3), then

$$\phi_u = 2\lambda\sin\tfrac{1}{2}\phi \quad \text{and} \quad \phi_v = 2\lambda^{-1}\sin\tfrac{1}{2}\phi.$$

Therefore

$$2\lambda u = \int^{\phi} \frac{d\phi}{\sin\frac{1}{2}\phi}$$

$$= 2\ln(\tan\tfrac{1}{4}\phi) + f(v),$$

where f is a function of integration. Similarly,

$$2\lambda^{-1}v = 2\ln(\tan\tfrac{1}{4}\phi) + g(u).$$

Therefore

$$\tan\tfrac{1}{4}\phi = C\exp(\lambda u + v/\lambda)$$

for some constant C. Thus the trivial solution ψ_o of the sine-Gordon equation (1) has generated the solution ϕ_1 of equation (4), where

$$\phi_1(u,v) = 4\tan^{-1}\{C\exp(\lambda u + v/\lambda)\}. \tag{5}$$

Of course, equations (1) and (4) are the same, so that ψ_o has generated another solution of the sine-Gordon equation.

Reverting to the sine-Gordon equation in its original form,

$$\psi_{tt} - \psi_{xx} + \sin\psi = 0, \tag{6}$$

with $u = \frac{1}{2}(x - t)$ and $v = \frac{1}{2}(x + t)$, we find that

$$\phi_1 = 4\tan^{-1}[C\exp\{(x - Ut)/\sqrt{1 - U^2}\}], \tag{7}$$

where

$$U = (\lambda^2 - 1)/(\lambda^2 + 1).$$

It can now be seen that we have generated the kink solution of §6.2 *Case* 5.

This method may be repeated, so that ϕ_1 generates a new solution ψ_2. To do this one puts $\phi = \phi_1$ into the Bäcklund transformation, (2) and (3), and eventually finds a solution representing the interaction of two solitons. Thus one can find the solution (6.4.3). The solution may be repeated further to find a solution representing the interaction of many solitons.

3 *The KdV equation*

Miura's transformation (Problem 4.3) may be written with the modified KdV equation,

$$v_x = u - v^2 \tag{1}$$

and

$$v_t = 6v^2v_x - v_{xxx}, \tag{2}$$

to imply that u satisfies the KdV equation,

$$P(u) = 0, \tag{3}$$

where $P(u) = u_t - 6uu_x + u_{xxx}$. Note that we can eliminate higher derivatives of v from equation (2) by use of equation (1), so that equations (1) and (2) are essentially a Bäcklund transformation.

Also the consistency condition, $\psi_{xxt} = \psi_{txx}$, of the scattering problem itself,

$$\psi_{xx} = (u + \kappa^2)\psi \tag{4}$$

and

$$\psi_t = -4\psi_{xxx} + 6u\psi_x + 3u_x\psi - 4\kappa^3\psi, \tag{5}$$

implies that $P(u) = 0$.

Another approach is due to Wahlquist & Estabrook (1973). Following them, we first transform the dependent variable of the KdV equation and then use a Bäcklund transformation. So we define a new dependent variable w by

$$w_x = u \tag{6}$$

and the operator Q by $Q(w) = w_t - 3w_x^2 + w_{xxx}$. Therefore

$$\{Q(w)\}_x = P(u)$$

identically. It follows that if u satisfies the KdV equation (3) then

$$Q(w) = f(t)$$

for some function f of integration. We may take $f = 0$ without loss of generality, replacing w by $w - \int^t f$ if necessary. Then

$$Q(w) = 0. \tag{7}$$

Now consider the Bäcklund transformation,

$$w_x + w_x' = 2\lambda + \tfrac{1}{2}(w - w')^2 \tag{8}$$

and

$$w_t + w_t' = -(w - w')(w_{xx} - w_{xx}') + 2(u^2 + uu' + u'^2), \tag{9}$$

where w and w' correspond to u and u' respectively, i.e. where $w_x = u$ and $w_x' = u'$, and λ is a real parameter. On assuming the transformation, we may take $\partial^2(8)/\partial x^2 + (9)$ to deduce that

$$Q(w) + Q(w') = 0,$$

and may take $\partial(9)/\partial x - \partial(8)/\partial t$ together with $\partial(8)/\partial x$ to deduce that

$$(w - w')\{Q(w) - Q(w')\} = 0\,.$$

Therefore equations (8) and (9) imply that w and w' each satisfy equation (7) and thence that u and u' each satisfy equation (3).

To use the Bäcklund transformation, (8) and (9), to generate solutions for the interaction of solitons, start with the trivial solution $u' = u'_o$ of the KdV equation, namely $u'_o = 0$. This corresponds to the solution $w' = w'_o = 0$ of equation (7). Then the transformation gives

$$w_x = 2\lambda + \tfrac{1}{2}w^2$$

and

$$w_t = - ww_{xx} + 2u^2.$$

The first of these equations may be integrated to give

$$w(x,t) = - 2\kappa\tanh\{\kappa x - f(t)\}$$

if $\lambda < 0$ and $\kappa = (-\lambda)^{\frac{1}{2}}$, where f is a function of integration. Also $w_{xx} = ww_x$. Therefore the second equation of the transformation gives

$$w_t = - w\, w_{xx} + 2w_x^2 = 2w_x(w_x - \tfrac{1}{2}w^2)$$

$$= 4\lambda w_x.$$

Therefore

$$\dot{f} = 4\lambda\kappa = - 4\kappa^3$$

and $w = w_1$, where

$$w_1(x,t) = - 2\kappa\tanh\{\kappa(x - x_o - 4\kappa^2 t)\} \qquad (10)$$

for $|w_1| < 2\kappa$. This gives the well-known solution for the solitary wave,

$$u_1(x,t) = w_{1x}(x,t) = - 2\kappa^2\operatorname{sech}^2\{\kappa(x - x_o - 4\kappa^2 t)\}. \qquad (11)$$

There is also a singular solution

$$w_1(x,t) = - 2\kappa\coth\{\kappa(x - x_o - 4\kappa^2 t)\} \qquad (12)$$

for $|w_1| > 2\kappa$.

There is an elegant way to repeat this process and thus generate solutions for interactions of many solitons. First suppose that two solutions, w_1 and w_2, are obtained from the Bäcklund transformation, (8) and (9), and the same given solution w_o (not necessarily the trivial solution) of equation (7) with two different values, λ_1 and λ_2, of λ. Thus, in particular, we suppose that

$$w_{1x} + w_{ox} = -2\lambda_1 + \tfrac{1}{2}(w_1 - w_o)^2 \tag{13}$$

and

$$w_{2x} + w_{ox} = -2\lambda_2 + \tfrac{1}{2}(w_2 - w_o)^2. \tag{14}$$

Similarly, we can construct a solution w_{12} from w_1 and λ_2 and a solution w_{21} from w_2 and λ_1, with

$$w_{12x} + w_{1x} = -2\lambda_2 + \tfrac{1}{2}(w_{12} - w_1)^2 \tag{15}$$

and

$$w_{21x} + w_{2x} = -2\lambda_1 + \tfrac{1}{2}(w_{21} - w_2)^2. \tag{16}$$

Now Bianchi's *theorem of permutability* gives

$$w_{21} = w_{12} \tag{17}$$

identically. (This important theorem is widely applicable (cf. Eisenhart 1909, p.286) and is very useful in iteration of multisoliton solutions, as is illustrated by the present example.) Therefore equations (13) - (14) - (15) + (16) give

$$0 = -4(\lambda_1 - \lambda_2) + \tfrac{1}{2}(w_1 - w_o)^2 - \tfrac{1}{2}(w_2 - w_o)^2 - \tfrac{1}{2}(w_{12} - w_1)^2$$
$$+ \tfrac{1}{2}(w_{12} - w_2)^2.$$

Therefore

$$w_{12} = w_o + \frac{4(\lambda_1 - \lambda_2)}{w_1 - w_2} , \tag{18}$$

giving w_{12} in simple *algebraic* terms of λ_1, λ_2, w_o, w_1 and w_2. Thus generation of solutions is remarkably simple, although not entirely free of tedium.

For an example, take $w_o = 0$, $\lambda_1 = -1$, $w_1 = -2\tanh(x - 4t)$, $\lambda_2 = -4$ and $w_2 = -4\coth(2x - 32t)$. Then

$$w_{12}(x,t) = -12/\{2\coth(2x - 32t) - \tanh(x - 4t)\}. \tag{19}$$

Therefore

$$u_{12}(x,t) = -12 \frac{3 + 4\cosh(2x - 8t) + \cosh(4x - 64t)}{\{3\cosh(x - 28t) + \cosh(3x - 36t)\}^2} \tag{20}$$

in agreement with the solution (4.10.5) found for the interaction of two solitons. This method is patently easier than the one in §4.10, and may be used to generate the solution for the interaction of three or more solitons.

Problems

7.1 *The Cauchy-Riemann relations.* Show that if

$$v_x = -u_y \quad \text{and} \quad u_x = v_y$$

then

$$\nabla^2 u = 0 \quad \text{and} \quad \nabla^2 v = 0.$$

7.2 *The Burgers equation.* Use the Bäcklund transformation,

$$v_x = -uv/2\nu$$

and

$$v_t = (u^2 - 2\nu u_x)v/4\nu,$$

to deduce that

$$u_t + uu_x = \nu u_{xx}$$

and

$$v_t = \nu v_{xx}.$$

[Cf. Problem 4.2; Forsyth (1906, p.101).]

7.3 *A modified KdV equation.* Use the Bäcklund transformation,

$$v_x = u - v^2 \quad \text{and} \quad v_t = -u_{xx} + 2(uv_x + u_x v),$$

to deduce that

$$u_t - 6uu_x + u_{xxx} = 0 \quad \text{and} \quad v_t - 6v^2 v_x + v_{xxx} = 0.$$

Defining θ such that $v = \theta_x/\theta$ in order to solve the above Riccati equation for v, show that

$$\theta_{xx} - u\theta = 0.$$

Taking the (singular) solution $u(x,t) = 2\operatorname{cosech}^2(x - 4t)$ of the KdV equation, deduce that

$$\{(1 - T^2)\theta_T\}_T - 2T^{-2}\theta = 0,$$

where the new independent variable is defined as $T = \tanh(x - 4t)$. Hence show that θ is in general a linear combination of T^{-1} and $(2T)^{-1}\ell n\{(1 + T)/(1 - T)\} - 1$. Taking $\theta = (2T)^{-1}\ell n\{(1 + T)/(1 - T)\} - 1$, conclude that $v(x,t) = \{T - (1 - T^2)(x - 4t)\}/T(x - 4t - T)$ is a (singular) solution of the modified KdV equation.

[Cf. Miura's transformation in Problem 4.3.]

7.4 *Another modified KdV equation.* Let $u = w_x$, and define the operators P and Q such that $P(u) = u_t + 6u^2u_x + u_{xxx}$ and $Q(w) = w_t + 2w_x^3 + w_{xxx}$. Then show that $P(u) = \{Q(w)\}_x$.

Use the Bäcklund transformation,

$$w'_x = - w_x + 2k\sin(w' - w)$$

and

$$w'_t = - w_t - 2k\{(u'_x - u_x)\cos(w' - w) + (u'^2 + u^2)\sin(w' - w)\},$$

where $u' = w'_x$ and k is a parameter, to show that $Q(w) = 0$ and $Q(w') = 0$ and thence that $P(u) = 0$ and $P(u') = 0$.

Show that if $w = 0$, then

$$w'(x,t) = \pm\, 2\tan^{-1}[\exp\{2k(x - x_0 - 4k^2t)\}].$$

Given w, k_1 and k_2, consider the construction of functions w_1, w_2, w_{12} and w_{21} such that

$$w_{1x} + w_x = 2k_1\sin(w_1 - w),$$

$$w_{2x} + w_x = 2k_2\sin(w_2 - w),$$

$$w_{12x} + w_{1x} = 2k_2 \sin(w_{12} - w_1)$$

and

$$w_{21x} + w_{2x} = 2k_1 \sin(w_{21} - w_2).$$

Using the theorem of permutability that $w_{21} = w_{12}$, deduce that

$$w_{12} = w + 2\tan^{-1}\left\{\frac{k_1 + k_2}{k_1 - k_2} \tan\tfrac{1}{2}(w_2 - w_1)\right\}.$$

[Wadati (1974), Lamb (1980, p.257).]

CHAPTER 8 EPILOGUE

1 *Epilogue*

The essence of this book is the description of the method of inverse scattering. The book is too short to do more than outline the chief properties of solitons and indicate some lesser properties by a few remarks and problems. For that reason the chief properties stand out more clearly.

Research into the physical, earth and life sciences has led to the study of hundreds of nonlinear equations. We have mentioned only a little of this. Indeed, it is too wide to be described in any single volume. Each reader of this introduction to solitons may, however, go on to study the derivation of the equations used in his own specialist field. On having obtained an appropriate nonlinear system, it is natural to seek the waves of permanent form and to test whether they are stable and whether they are solitons. The existence of soliton interactions is the exception rather than the rule. Yet scores of nonlinear systems are already known to have soliton solutions. Even now it is not known how to ascertain definitely whether any given nonlinear system has solitons which may preserve their identities after interacting with one another. The preceding chapters, nonetheless, outline many ideas which may determine, or at least suggest, that a given nonlinear system has, or has not, soliton solutions and which indicate how the properties of the solitons may be found. The ideas are summarized in the following points.

(i) First seek steady-state solutions, i.e. waves of permanent form, as described in Chapter 2. A soliton is such a solution, although such a solution is in general not a soliton. These special solutions of the nonlinear system are likely to be important in many ways.

(ii) Another important class of special solutions are the similarity solutions. They may be found by use of dimensional analysis or of the

group of transformations from one dynamically similar solution to another.
(iii) This group and other groups of transformations under which the nonlinear system is invariant are likely to underlie the character of all methods of solution of the system. In addition to the group of transformations among dynamically similar solutions, we have met examples of translations, Galilean transformations and various time and space symmetries. Recognition of those groups may help the solver of the system. Problem 4.3 gives an historic example of this.

(iv) Seek as many conservation laws of the given system as possible. An infinity of conservation laws seems to be associated with the existence of soliton interactions, and also arises from the time-independence of the transmission coefficient of short waves in an associated scattering problem (as in Problem 4.8). The integrals of the conserved densities emerge thus as constants of the motion.

(v) The crucial task is the search for the associated scattering problem whose eigenvalues are constants when the given nonlinear system is satisfied. The Lax method (described in Chapter 5) and the two-component method (in §6.6) show how such scattering problems may be found. There are, in addition, powerful generalizations of the two-component method and a few other methods of inverse scattering. However, the identification of a scattering problem appropriate to a given nonlinear system may be said to be still more an art than a science, because it depends upon the use of trial and error rather than an algorithm. Once the scattering problem is identified, the evolution equation for the eigenfunctions follows in a relatively straightforward way.

(vi) Seek a relevant Bäcklund transformation, as described in Chapter 7, if the given nonlinear system is a partial differential one. Bäcklund transformations were shown to be closely associated with the method of inverse scattering, and to be useful in finding multisoliton solutions.

(vii) It is also useful to find properties of the singularities of the ordinary differential equations governing waves of permanent form, and similarity solutions. In particular, the *Painlevé property* of an equation that each movable singularity is a pole (Ince 1927, §14.4) seems to be associated with the existence of solitons (Ablowitz, Ramani & Segur 1980). A similar property of the singularities of the original nonlinear partial differential system also seems to determine whether solitons exist (Weiss, Tabor & Carnevale 1983). This property is possessed by Painlevé transcendents (which arose in Problems 2.4 and 6.6).

(viii) Seek a Hamiltonian representation of the given nonlinear system. Of course, a partial differential system would have an infinite-dimensional Hamiltonian representation or none at all. (Examples of Hamiltonians were given in Problems 3.6 and 6.4.) In fact, the method of inverse scattering is effectively a canonical transformation of the variables of a completely integrable Hamiltonian system to action-angle variables. This emphasizes the special nature of a nonlinear system which is solvable by an inverse scattering transform.

Modern research into solitons started with Zabusky & Kruskal's (1965) numerical solution of the KdV equation with periodic boundary conditions (described in §1.3). Periodic boundary conditions are more difficult to treat mathematically than the ones at infinity which have been taken in this book and which have been mostly considered by research workers. It should be noted, however, that in spite of this concentration upon conditions at infinity there has been some success in applying the method of inverse scattering to problems with periodic boundary conditions.

We have mentioned only two systems (Problems 2.8 and 2.14) with more than one space variable. Derrick (1964) showed that no steady solution of a certain class of scalar wave equation is stable in space of higher dimension than one, but his assumptions restrict the application of his results. Indeed, solitons have an even richer structure in spaces of higher dimension than in a one-dimensional space. Perhaps the most important problem of higher spatial dimension is the one of elementary particles in quantum field theory.

The superficial similarity between the properties of solitons and of elementary particles is striking. Solitons may propagate without change of form. A soliton may be regarded as a local confinement of the energy of the wave field. When two solitons collide, each may come away with the same character as it had before the collision. When a soliton meets an antisoliton, both may be annihilated. Elementary particles share these properties. So if an appropriate system of nonlinear field equations admits soliton solutions then these solitons may represent elementary particles and have properties which may be confirmed by observations of particles. Quantum field theories are very complicated, so research into their soliton solutions, although intense now, is at a much more primitive stage than that into the simple soliton solutions of the KdV and sine-

Gordon equations. In particular, the Yang-Mills field equations seem to be a fruitful model unifying electromagnetic and weak forces. They admit solutions localized in space which represent very heavy elementary particles. They also have solutions, called instantons, localized in time as well as space; these are interpreted as quantum-mechanical transitions between different states of a particle. A popular introduction to current research in all this has been written by Rebbi (1979), opening up a fascinating vista.

*APPENDIX A DERIVATION OF THE INTEGRAL EQUATION FOR INVERSE SCATTERING

When the scattering problem was inverted in §4.6, we merely quoted the result which we needed to use. A derivation of this result is sketched here.

Gel'fand & Levitan (1951) solved the problem of inverse scattering in one dimension, but their motivation is unconnected with our interest in the solution of initial-value problems for nonlinear equations. They determined the scattering potential u by use of the *spectral function*. This is a monotonic increasing function $\rho(\lambda)$ which is constant for $\lambda < 0$ except for jumps of c_n at the points $\lambda = -\kappa_n^2$ belonging to the discrete spectrum and whose increase for $\lambda > 0$ is also specified by the scattering problem, (4.3.1) and (4.3.2). The integral equation in the form (4.6.2) is due to Marchenko (1955) and Kay & Moses (1956a). It is more easily derived by a method of Balanis (1972), which is sketched below.

Accordingly, regard the scattering equation (4.3.1),

$$\psi_{xx} + (k^2 - u)\psi = 0, \tag{1}$$

as the Fourier transform with respect to τ of the equation,

$$\phi_{xx} - \phi_{\tau\tau} - u\phi = 0, \tag{2}$$

where $\phi(x,\tau,t) = \int_{-\infty}^{\infty} \psi(x,k,t)e^{-ik\tau}dk$. Next regard equation (2) in turn as a time-dependent wave equation with given potential u, although the 'time' τ is not the parameter t, which will be ignored for the moment.

We then may consider a solution representing a given incident wave coming from $x = \infty$ with a transmitted wave at $x = -\infty$ and a reflected wave at $x = \infty$. In particular, take the incident wave to be a pulse and the reflected wave to be $B(x - \tau)$ so that

$$\phi(x,\tau) \sim \phi_\infty(x,\tau) \quad \text{as} \quad x \to \infty, \tag{3}$$

where

$$\phi_\infty(x,\tau) = \delta(x + \tau) + B(x - \tau). \tag{4}$$

This solution is valid as $x \to \infty$ because $u \to 0$. The corresponding complete solution of equation (2) is in fact of the form

$$\phi(x,\tau) = \phi_\infty(x,\tau) + \int_x^\infty K(x,\xi)\phi_\infty(\xi,\tau)d\xi. \tag{5}$$

This solution may be verified by substitution into equation (2) if

$$K_{\xi\xi} - K_{xx} + u(x)K = 0 \quad \text{for} \quad \xi > x, \tag{6}$$

$$u(x) = -2\,\frac{dK(x,x)}{dx} \tag{7}$$

and

$$K, K_\xi \to 0 \quad \text{as} \quad \xi \to \infty. \tag{8}$$

By the standard theory of wave equations, this problem, (6),(7) and (8), is well posed and has a unique solution K.

Now waves governed by equation (2) travel with unit speed, so the property of causality gives

$$\phi(x,\tau) = 0 \quad \text{if} \quad x + \tau < 0,$$

i.e.

$$\phi_\infty(x,\tau) + \int_x^\infty K(x,\tau)\phi_\infty(\xi,\tau)d\xi = 0 \quad \text{if} \quad x + \tau < 0,$$

i.e.

$$B(x - \tau) + K(x - \tau) + \int_x^\infty K(x,\xi)B(\xi - \tau)d\xi = 0. \tag{9}$$

This integral equation takes the form of equation (4.6.2) on putting

$\tau = -y$.

It only remains to evaluate B in terms of the scattering data. Now $\delta(x + \tau) = (2\pi)^{-1}\int_{-\infty}^{\infty} e^{-ik(x+\tau)}dk$, so Fourier analysis of equation (2) and standard properties of direct scattering theory may be used to show that B is as given by equation (4.6.3), where c_n and b are defined in §§4.5 and 4.3 respectively.

Whitham (1974, pp.590-2) has extended Balanis's method to find, in an interesting and simple way, the solution of the initial-value problem of the KdV equation itself.

BIBLIOGRAPHY AND AUTHOR INDEX

The numbers in square brackets following each entry give the pages of this book on which the entry is cited.

Ablowitz, M.J., Kaup, D.J., Newell, A.C. & Segur, H. 1973 Method for solving the sine-Gordon equation. Phys. Rev. Lett. 30, 1262-4. [42, 99-100, 106]

Ablowitz, M.J., Kaup, D.J., Newell, A.C. & Segur, H. 1974 The inverse scattering transform – Fourier analysis for nonlinear problems. Studies Appl. Math. 53, 249-315. [100]

Ablowitz, M.J., Ramani, A. & Segur, H. 1980 A connection between nonlinear evolution equations and ordinary differential equations of P-type. I. J. Math. Phys. 21, 715-21. [122]

Abramowitz, M. & Stegun, I.A. (Editors) 1964 Handbook of Mathematical Functions. Washington, D.C.: Nat. Bureau of Standards. Also New York: Dover (1965). [18, 19, 56, 58]

Aris, R. 1975 The Mathematical Theory of Diffusion and Reaction in Permeable Catalysts. Oxford University Press. 2 vols. [28]

Balanis, G.N. 1972 The plasma inverse problem. J. Math. Phys. 13, 1001-5. [125]

Barbashov, B.M. & Chernikov, N.A. 1967 Solution of the two plane wave scattering problems in a nonlinear scalar field theory. Sov. Phys. JETP 24, 437-42 (in English transl.). [39]

Benjamin, T.B. 1967 Internal waves of permanent form in fluids of great depth. J. Fluid Mech. 29, 559-92. [11, 26]

Benjamin, T.B. 1972 The stability of solitary waves. Proc. Roy. Soc. A 328, 153-83. [17]

Benjamin, T.B., Bona, J.L. & Mahony, J.J. 1972 Model equations for long waves in nonlinear dispersive systems. Phil. Trans. A 272, 47-78. [11]

Benney, D.J. 1966 Long non-linear waves in fluid flows. J. Math. & Phys. 45, 52-63. [8]

Bluman, G.W. & Cole, J.D. 1974 Similarity Methods for Differential Equations. New York: Springer-Verlag. [4]

Bona, J.L. & Smith, R. 1975 The initial-value problem for the Korteweg-de Vries equation. Phil. Trans. A 278, 555-604. [40]

Born, M. & Infeld, L. 1934 Foundations of the new field theory. Proc. Roy. Soc. A 144, 425-51. [39]

Boussinesq, J. 1871 Théorie de l'intumescence liquide appelée onde solitaire ou de translation, se propageant dans un canal rectangulaire. Compte Rendus Acad. Sci. Paris 72, 755-9. [2]

Burgers, J.M. 1948 A mathematical model illustrating the theory of turbulence. Adv. Appl. Mech. 1, 171-99. [29, 71]

Clairin, J. 1903 Sur quelques équations aux derivées partielles du second ordre. Ann. Fac. Sci. Univ. Toulouse (2) 5, 437-58. [111]

Cole, J.D. 1951 On a quasi-linear parabolic equation occurring in aerodynamics. Quart. Appl. Math. 9, 225-36. [71]

Coleman, S. 1975 Quantum sine-Gordon equation as the massive Thirring model. Phys. Rev. D 11, 2088-97. [88]

Davey, A. 1972 The propagation of a weak nonlinear wave. J. Fluid Mech. 53, 769-81. [12]

Davis, R.E. & Acrivos, A. 1967 Solitary internal waves in deep water. J. Fluid Mech. 29, 593-607. [26]

Derrick, G.H. 1964 Comments on nonlinear wave equations as models for elementary particles. J. Math. Phys. 5, 1252-4. [123]

Drazin, P.G. 1963 On one-dimensional propagation of long waves. Proc. Roy. Soc. A 273, 400-11. [73, 74]

Drazin, P.G. 1977 On the stability of cnoidal waves. Quart. J. Mech. Appl. Math. 30, 91-105. [17]

Dryuma, V.S. 1974 Analytic solution of the two-dimensional Korteweg-de Vries (KdV) equation. Sov. Phys. JETP Lett. 19, 387-8. [86]

Eisenhart, L.P. 1909 A Treatise on the Differential Geometry of Curves and Surfaces. Boston, Mass.: Ginn. Also New York: Dover (1960). [87, 108, 111, 116]

Faddeev, L.D. 1958 On the relation between the S-matrix and potential for the one-dimensional Schrödinger operator. Dokl. Akad. Nauk SSSR 121, 63-6. An expanded English version is in Amer. Math. Soc. Transl. (2) 65 (1967), 139-66. [40]

Fife, P.C. 1979 Mathematical Aspects of Reacting and Diffusing Systems. Springer Lecture Notes in Biomath., no. 28. [28]

Fisher, R.A. 1937 The wave of advance of advantageous genes. Ann. Eugenics 7, 355-69. Also Collected Papers, vol. IV, pp. 69-83. University of Adelaide. [28]

Forsyth, A.R. 1906 Theory of Differential Equations. Part IV. Partial Differential Equations. Vol. VI. Cambridge University Press. Also New York: Dover (1959). [71, 108, 111, 118]

Freeman, N.C. 1980 Soliton interactions in two-dimensions. Adv. Appl. Mech. 20, 1-37. [26]

Frenkel, J. & Kontorova, T. 1939 On the theory of plastic deformation and twinning. Fiz. Zhurnal 1, 137-49. [87]

Gardner, C.S. 1971 Korteweg-de Vries equation and generalizations. IV. The Korteweg-de Vries equation as a Hamiltonian system. J. Math. Phys. 12, 1548-51. [38]

Gardner, C.S., Greene, J.M., Kruskal, M.D. & Miura, R.M. 1967 Method for solving the Korteweg-de Vries equation. Phys. Rev. Lett. 19, 1095-7. [42, 71]

Gardner, C.S., Greene, J.M., Kruskal, M.D. & Miura, R.M. 1974 Korteweg-de Vries equation and generalizations. VI. Methods for exact solution. Comm. Pure Appl. Math. 27, 97-133. [51, 70, 77]

Gel'fand, I.M. & Fomin, S.V. 1963 Calculus of Variations. Englewood Cliffs, N.J.: Prentice Hall. [33]

Gel'fand, I.M. & Levitan, B.M. 1951 On the determination of a differential equation from its spectral function. Izv. Akad. Nauk SSSR Ser. Mat. 15, 309-66 (in Russian). Also in Amer. Math. Soc. Transl. (2) 1 (1955), 253-304 (in English). [50, 125]

Gibbon, J.D., Caudrey, P.J., Bullough, R.K. & Eilbeck, J.C. 1973 An N-soliton solution of a nonlinear optics equation derived by a general inverse method. Lett. Nuovo Cimento 8, 775-9. [107]

Gordon, W. 1926 Der Comptoneffekt der Schrödingerschen Theorie. Zeit. für Phys. 40, 117-33. [87]
Greig, I.S. & Morris, J.L. 1976 A hopscotch method for the Korteweg-de Vries equation. J. Comp. Phys. 20, 64-80. [67]
Griffel, D.H. 1981 Applied Functional Analysis. Chichester: Ellis Horwood. [79]
Grindlay, J. & Opie, A.H. 1977 The non-linear and dispersive elastic solid: standing waves. J. Phys. C: Solid State Phys. 10, 947-54. [25]
Hasimoto, H. & Ono, H. 1972 Nonlinear modulation of gravity waves. J. Phys. Soc. Japan 33, 805-11. [12]
Hirota, R. 1971 Exact solution of the Korteweg-de Vries equation for multiple collisions of solitons. Phys. Rev. Lett. 27, 1192-4. [78]
Hirota, R. 1976 Direct method of finding exact solutions of nonlinear evolution equations. In Bäcklund Transformations, ed. R.M. Miura, Springer Lecture Notes in Math., no. 515, pp. 40-68. [78]
Hopf, E. 1950 The partial differential equation $u_t + uu_x = \mu u_{xx}$. Comm. Pure Appl. Math. 3, 201-30. [71]
Ince, E.L. 1927 Ordinary Differential Equations. London: Longmans, Green. Also New York: Dover (1956). [24, 104, 122]
Kadomtsev, B.B. & Petviashvili, V.I. 1970 On the stability of solitary waves in weakly dispersing media. Dokl. Akad. Nauk SSSR 192, 753-6. Also in Sov. Phys. Dokl. 15, 539-41 (in English transl.). [26]
Kay, I. & Moses, H.E. 1956a The determination of the scattering potential from the spectral measure function. III. Calculation of the scattering potential from the scattering operator for the one-dimensional Schrödinger equation. Nuovo Cimento (10) 3, 276-304. [50, 125]
Kay, I. & Moses, H.E. 1956b Reflectionless transmission through dielectrics and scattering potentials. J. Appl. Phys. 27, 1503-8. [68]
Kelley, P.L. 1965 Self-focusing of optical beams. Phys. Rev. Lett. 15, 1005-8. [12]
Kittel, C. 1976 Introduction to Solid State Physics. 5th edn. New York: Wiley. [28]
Klein, O. 1927 Elektrodynamik und Wellenmechanik vom Standpunkt des Korrespondenzprinzips. Zeit. für Phys. 41, 407-42. [87]
Kolmogorov, A.N., Petrovsky, I.G. & Piscounov, N.S. 1937 Study of the diffusion equation for growth of a quantity of a substance and application to a biological problem. Bull. Moscow State University A1 (6), (in Russian) 26 pp. [28]
Korteweg, D.J. & de Vries, G. 1895 On the change of form of long waves advancing in a rectangular canal, and on a new type of long stationary waves. Phil. Mag. (5) 39, 422-43. [3, 8, 17]
Lamb, G.L., Jr 1970 Higher conservation laws in ultrashort optical pulse propagation. Phys. Lett. 32A, 251-2. [103]
Lamb, G.L., Jr 1971 Analytical description of ultrashort optical pulse propagation in a resonant medium. Rev. Modern Phys. 43, 99-124. [107]
Lamb, G.L., Jr 1980 Elements of Soliton Theory. New York: Wiley-Interscience. [120]
Landau, L.D. & Lifshitz, E.M. 1959 Fluid Mechanics. London: Pergamon. [6]
Landau, L.D. & Lifshitz, E.M. 1965 Quantum Mechanics Non-relativistic Theory. 2nd edn. London: Pergamon. [43, 73]

Larichev, V.D. & Reznik, G.M. 1976 Two-dimensional Rossby soliton: an exact solution. Rep. U.S.S.R. Acad. Sci 231 (5) (in Russian). Also in POLYMODE News 19, pp.3 & 6 (in English). [30]

Lax, P.D. 1968 Integrals of nonlinear equations of evolution and solitary waves. Comm. Pure Appl. Math. 21, 467-90. [42, 79, 83, 84]

Lighthill, J. 1978 Waves in Fluids. Cambridge University Press. [5]

Liouville, J. 1853 Sur l'équation aux différences partielles $d^2\log\lambda/dudv \pm \lambda/2a^2 = 0$. J. Math. Pures Appl. 18, 71-2. [110]

McWilliams, J.C. & Zabusky, N.J. 1982 Interactions of isolated vortices I: Modons colliding with modons. Geophys. Astrophys. Fluid Dyn. 19, 207-27. [30]

Marchenko, V.A. 1955 Dokl. Akad. Nauk SSSR 104, 695-8 (in Russian). [50, 125]

Miles, J.W. 1978 On the evolution of a solitary wave for very weak nonlinearity. J. Fluid Mech. 87, 773-83. [73]

Miura, R.M. 1968 Korteweg-de Vries equation and generalizations. I. A remarkable explicit nonlinear transformation. J. Math. Phys. 9, 1202-4. [71]

Miura, R.M. 1974 The Korteweg-de Vries equation: a model equation for nonlinear dispersive waves. In Nonlinear Waves, eds. S. Leibovich & A.R. Seebass, pp. 212-34. Ithaca, N.Y.: Cornell University Press. [37]

Miura, R.M. 1976 The Korteweg-de Vries equation: a survey of results. SIAM Rev. 18, 412-59. [10, 76]

Miura, R.M., Gardner, C.S. & Kruskal, M.D. 1968 Korteweg-de Vries equation and generalizations. II. Existence of conservation laws and constants of motion. J. Math. Phys. 9, 1204-9. [34, 37, 75]

Morse, P.M. & Feshbach, H. 1953 Methods of Theoretical Physics. New York: McGraw-Hill. Part II (i.e. vol. II). [74]

Olver, P.J. 1979 Euler operators and conservation laws of the BBM equation. Math. Proc. Camb. Phil. Soc. 85, 143-60. [37]

Ono, H. 1975 Algebraic solitary waves in stratified fluids. J. Phys. Soc. Japan 39, 1082-91. [26, 38]

Peregrine, D.H. 1966 Calculations of the development of an undular bore. J. Fluid Mech. 25, 321-30. [11]

Perring, J.K. & Skyrme, T.H.R. 1962 A model unified field equation. Nuclear Phys. 31, 550-5. [88, 95]

Rayleigh, Lord 1876 On waves. Phil. Mag. (5) 1, 257-79. Also Sci. Papers, vol. I, pp.251-71. Cambridge University Press. [2]

Rebbi, C. 1979 Solitons. Sci. American 240 (2), 76-91. [124]

Russell, J.S. 1844 Report on waves. Rep. 14th Meet. Brit. Assoc. Adv. Sci. York 1844, 311-90. [1, 2, 3]

Scott, A.C. 1970 Active and Nonlinear Wave Propagation in Electronics. New York: Wiley-Interscience. [101]

Segur, H. 1973 The Korteweg-de Vries equation and water waves. Solutions of the equation. Part 1, J. Fluid Mech. 59, 721-36. [52]

Stern, M.E. 1975 Minimal properties of planetary eddies. J. Marine Res. 33, 1-13. [30]

Taylor, G.I. 1910 The conditions necessary for discontinuous motion in gases. Proc. Roy. Soc. A 84, 371-7. Also Sci. Papers, vol. III, pp. 1-6. Cambridge University Press. [29]

Titchmarsh, E.C. 1948 Introduction to the Theory of Fourier Integrals. 2nd edn. Oxford University Press. [11, 26]

Toda, M. 1967a Vibration of a chain with nonlinear interaction. J. Phys. Soc. Japan 22, 431-6. [28]

Toda, M. 1967b Wave propagation in anharmonic lattices. J. Phys. Soc. Japan 23, 501-6. [25, 28]
Wadati, M. 1974 Bäcklund transformation for solutions of the modified Korteweg-de Vries equation. J. Phys. Soc. Japan 36, 1498. [120]
Wadati, M. & Toda, M. 1972 The exact N-soliton solution of the Korteweg-de Vries equation. J. Phys. Soc. Japan 32, 1403-11. [51, 70]
Wahlquist, H.D. & Estabrook, F.B. 1973 Bäcklund transformation for solutions of the Korteweg-de Vries equation. Phys. Rev. Lett. 31, 1386-90. [113]
Washimi, H. & Taniuti, T. 1966 Propagation of ion-acoustic solitary waves of small amplitude. Phys. Rev. Lett. 17, 996-8. [8]
Weiss, J., Tabor, M. & Carnevale, G. 1983 The Painlevé property for partial differential equations. J. Math. Phys. 24 (in press). [122]
Whitham, G.B. 1967 Variational methods and applications to water waves. Proc. Roy. Soc. A 299, 6-25. [11]
Whitham, G.B. 1974 Linear and Nonlinear Waves. New York: Wiley. [6, 11, 102, 127]
Wijngaarden, L. van 1968 On the equations of motion for mixtures of liquid and gas bubbles. J. Fluid Mech. 33, 465-74. [8]
Zabusky, N.J. 1967 A synergetic approach to problems of nonlinear dispersive wave propagation and interaction. In Nonlinear Partial Differential Equations, ed. W.F. Ames, pp. 223-58. New York: Academic Press. [23, 24]
Zabusky, N.J. & Kruskal, M.D. 1965 Interactions of "solitons" in a collisonless plasma and the recurrence of initial states. Phys. Rev. Lett. 15, 240-3. [1, 7, 8, 123]
Zakharov, V.E. 1971 Kinetic equation for solitons. Sov. Phys. JETP 33, 538-41 (in English transl.). [51, 70]
Zakharov, V.E. & Shabat, A.B. 1972 Exact theory of two-dimensional self-focusing and one-dimensional self-modulation of waves in nonlinear media. Sov. Phys. JETP 34, 62-9 (in English transl.). [38, 86, 100, 106]
Zakharov, V.E. & Shabat, A.B. 1973 Interaction between solitons in a stable medium. Sov. Phys. JETP 37, 823-8 (in English transl.). [31]

MOTION PICTURE INDEX

Various films about solitons have been made. Some 16 mm ones, which may be loaned or bought, are listed below. It is very instructive to see, in particular, computer-made animations of solitons and their interactions.

Eilbeck, J.C. F1973 Solitons and Bions in Nonlinear Optics. Sound, colour, 6 min.

Eilbeck, J.C. F1975 Numerical Solutions of the Régularized Long-Wave Equation. Silent, colour, 3 min.

Eilbeck, J.C. F1977a Boomerons. Silent, colour, 6 min.

Eilbeck, J.C. F1977b Zoomerons. Silent, colour, 6 min.

Eilbeck, J.C. F1978 Two-dimensional Solitons. Silent, colour, 4 min.

Eilbeck, J.C. F1981 Nonlinear Evolution Equations. Sound, colour, 12 min. [9, 62, 71]

Eilbeck, J.C. F1982 Kink Interactions in the ϕ^4 Model. Silent, colour, 6 min.

Eilbeck, J.C. & Lomdahl, P.S. F1982 Sine-Gordon Solitons. Sound, colour, 14 min. [96, 101]

The above films may be bought from FR80 Ops., Rutherford Laboratory, Chilton, Didcot, Oxfordshire OX11 0QX, England.

Zabusky, N.J., Kruskal, M.D. & Deem, G.S. F1965 Formation, Propagation and Interaction of Solitons (Numerical Solutions of Differential Equations Describing Wave Motions in Nonlinear Dispersive Media). Silent, b/w, 35 min. [9, 71]

The above film may be loaned from Bell Telephone Laboratories, Inc. Film Library, Murray Hill, New Jersey 07971, U.S.A.

SUBJECT INDEX